Birkhäuser

Modern Birkhäuser Classics

Many of the original research and survey monographs in pure and applied mathematics published by Birkhäuser in recent decades have been groundbreaking and have come to be regarded as foundational to the subject. Through the MBC Series, a select number of these modern classics, entirely uncorrected, are being re-released in paperback (and as eBooks) to ensure that these treasures remain accessible to new generations of students, scholars, and researchers.

Ernest B. Vinberg

Linear Representations of Groups

Translated from the Russian by A. Iacob

Reprint of the 1989 Edition

Birkhäuser

E. B. Vinberg
Department of Mathematics (Algebra)
Moscow State University
119899, GSP Moscow
Russia
vinberg@zebra.ru

2010 Mathematics Subject Classification 20C15, 20G05, 22E47, 20-01, 22-01, 22E45, 22D10

ISBN 978-3-0348-0062-4 e-ISBN 978-3-0348-0063-1
DOI 10.1007/978-3-0348-0063-1

Originally published as Lineinye predstavleniya grupp by Nauka (Glav. Red. FML), Moscow 1985. Translated from the Russ. By A. Iacob and published as Volume 2 under the same title in the Basler Lehrbücher – Birkhäuser Advanced Texts series by Birkhäuser Verlag, Switzerland, ISBN 3-7643-2288-8
Reprinted 2010 by Springer Basel AG

Cover design: deblik, Berlin

Springer Basel AG is part of Springer Science+Business Media

www.birkhauser-science.com

Contents

Preface

This book gives an exposition of the fundamentals of the theory of linear representations of finite and compact groups, as well as elements of the theory of linear representations of Lie groups. As an application we derive the Laplace spherical functions. The book is based on lectures that I delivered in the framework of the experimental program at the Mathematics-Mechanics Faculty of Moscow State University and at the Faculty of Professional Skill Improvement. My aim has been to give as simple and detailed an account as possible of the problems considered. The book therefore makes no claim to completeness. Also, it can in no way give a representative picture of the modern state of the field under study as does, for example, the monograph of A. A. Kirillov [3].

For a more complete acquaintance with the theory of representations of finite groups we recommend the book of C. W. Curtis and I. Reiner [2], and for the theory of representations of Lie groups, that of M. A. Naimark [6].

Introduction

The theory of linear representations of groups is one of the most widely applied branches of algebra. Practically every time that groups are encountered, their linear representations play an important role. In the theory of groups itself, linear representations are an irreplaceable source of examples and a tool for investigating groups.

In the introduction we discuss some examples and en route we introduce a number of notions of representation theory.

0. Basic Notions

0.1. The exponential function

$$t \mapsto \mathrm{e}^{ta} \quad (t \in \mathbf{R})$$

is, for every fixed $a \in \mathbf{R}$, a homomorphism of the additive group $\mathbf{R}$ into the multiplicative group $\mathbf{R}^*$. Are there any other homomorphisms $f : \mathbf{R} \to \mathbf{R}^*$? Without attempting to answer this question in full generality, we require that the function f be differentiable. The condition that f be a homomorphism is written:

$$f(t + u) = f(t)f(u)$$

for all $t, u \in \mathbf{R}$. Differentiating with respect to u and putting $u = 0$ we get

$$f'(t) = f(t)a,$$

where $a = f'(0)$. The general solution of this differential equation is $f(t) = C\mathrm{e}^{ta}$, but the condition that f be a homomorphism forces $f(0) = 1$, whence $C = 1$. Thus, *every differentiable group homomorphism of* $\mathbf{R}$ *into* $\mathbf{R}^*$ *is an exponential function.* This is one of the reasons why exponential functions play such an important role in mathematics.

In solving systems of linear differential equations with constant coefficients one encounters the *matrix exponential function*

$$(1) \qquad t \mapsto \mathrm{e}^{tA} \quad (t \in \mathbf{R},\ A \in \mathrm{L}_n(\mathbf{R})).$$

E.B. Vinberg, *Linear Representations of Groups*, Modern Birkhäuser Classics,
DOI 10.1007/978-3-0348-0063-1_1, © Birkhäuser Verlag 1989

Recall that it is defined as the sum of the series

$$\sum_{k=0}^{\infty} \frac{t^k A^k}{k!}$$

or, alternatively, as the solution of the matrix differential equation

$$(2) \qquad F'(t) = F(t)A \qquad (F : \mathbf{R} \to \mathrm{L}_n(\mathbf{R}))$$

with initial condition $F(0) = E$. The exponential matrix function has the property that

$$\mathrm{e}^{(t+u)A} = \mathrm{e}^{tA}\mathrm{e}^{uA} \quad \text{for all } t, u \in \mathbf{R},$$

i.e., it is a *homomorphism* of the group $\mathbf{R}$ into the group $\mathrm{GL}_n(\mathbf{R})$. As above, one can show that *every differentiable group homomorphism of* $\mathbf{R}$ *into* $\mathrm{GL}_n(\mathbf{R})$ *has the form* (1). This is a generalization of the preceding result, since $\mathbf{R}^* = \mathrm{GL}_1(\mathbf{R})$.

Example. It follows from the addition formulas for trigonometric functions that

$$(3) \qquad \begin{pmatrix} \cos(t+u) & -\sin(t+u) \\ \sin(t+u) & \cos(t+u) \end{pmatrix} = \begin{pmatrix} \cos t & -\sin t \\ \sin t & \cos t \end{pmatrix} \begin{pmatrix} \cos u & -\sin u \\ \sin u & \cos u \end{pmatrix}.$$

This says that the map

$$(4) \qquad F : t \mapsto \begin{pmatrix} \cos t & -\sin t \\ \sin t & \cos t \end{pmatrix}$$

is a group homomorphism of $\mathbf{R}$ into $\mathrm{GL}_2(\mathbf{R})$, and hence it has the form (1). Here

$$A = F'(0) = \begin{pmatrix} 0 & -1 \\ 1 & 0 \end{pmatrix}.$$

0.2. Let S_n denote the symmetric group of degree n and let K be an arbitrary field. To each permutation $\sigma \in S_n$ we assign the matrix

$$(5) \qquad M(\sigma) = E_{\sigma(1),1} + \dots + E_{\sigma(n),n},$$

where E_{ij} designates the matrix whose (i,j)-entry is the identity element of the field K, while the remaining entries are zero.

We claim that for all $\sigma, \tau \in S_n$:

$$(6) \qquad M(\sigma\tau) = M(\sigma)M(\tau).$$

In fact, one readily verifies the following multiplication rule for the matrices E_{ij}:

$$E_{ij}E_{\ell k} = \begin{cases} E_{ik} & \text{if } j = \ell, \\ 0 & \text{if } j \neq \ell. \end{cases}$$

Using it we find that

$$\begin{aligned} M(\sigma)M(\tau) &= E_{\sigma\tau(1),\tau(1)}E_{\tau(1),1} + \ldots + E_{\sigma\tau(n),\tau(n)}E_{\tau(n),n} \\ &= E_{\sigma\tau(1),1} + \ldots + E_{\sigma\tau(n),n} = M(\sigma\tau), \end{aligned}$$

as claimed.

We remark that the matrices $M(\sigma)$ are nonsingular. More precisely,

$$\det M(\sigma) = \Pi(\sigma)$$

where $\Pi(\sigma) = \pm 1$ denotes the parity of the permutation σ. Equality (6) says that M is a group homomorphism of S_n into $\mathrm{GL}_n(K)$.

0.3. Definition. A MATRIX REPRESENTATION of the group G over the field K is a homomorphism

$$T: G \to \mathrm{GL}_n(K)$$

of G into the group $\mathrm{GL}_n(K)$ of nonsingular matrices of order n over K. The number n is called the DIMENSION of the representation T.

The word "representation" should suggest that once a matrix representation is given, the elements of the group can be viewed as matrices or, in other words, that there is an isomorphism of the given group with some group of matrices. In the cases where the group G has a rather complicated structure, it may very well turn out that such a representation is the only simple way of describing G. For instance, the group $\mathrm{GL}_n(K)$ is defined as a matrix group, i.e., the identity map $\mathrm{Id}: \mathrm{GL}_n(K) \to \mathrm{GL}_n(K)$ is a matrix representation of $\mathrm{GL}_n(K)$.

We should, however, emphasize from the very beginning that in reality a matrix representation does not always give an isomorphism between the group G and a subgroup of $\mathrm{GL}_n(K)$. The reason is that one and the same matrix might correspond to distinct elements of G. To the reader familiar with the general properties of homomorphisms it should be clear that this occurs if and only if there are elements of G different from its identity element which are mapped to the identity element of the group $\mathrm{GL}_n(K)$, i.e., to the identity matrix.

The set of all elements of G which are taken by T into the identity matrix is a normal subgroup of G, called the KERNEL of the representation T and denoted by $\mathrm{Ker}\, T$. For example, the kernel of the representation (4) of $\mathbf{R}$ consists of all numbers $2\pi k$ with $k \in \mathbf{Z}$.

If $\mathrm{Ker}\, T$ reduces to the identity element of G, the representation T is said to be FAITHFUL. In this case G is isomorphic to the subgroup $T(G)$ of $\mathrm{GL}_n(K)$.

The representation of the group S_n constructed above in 0.2 provides an example.

The other extreme case occurs when $\operatorname{Ker} T = G$, i.e., all elements of G are taken by T into the identity matrix. Such a T is called a TRIVIAL representation.

0.4. We next discuss the *geometric approach* to the notion of representation.

Every matrix A of order n with entries in the field K defines a linear transformation $x \mapsto Ax$ of the space K^n of column vectors. Moreover, this correspondence between matrices and linear transformations is bijective and turns the multiplication of matrices into the multiplication (composition) of linear transformations. In particular, the group $\mathrm{GL}_n(K)$ is canonically isomorphic to the group $\mathrm{GL}(K^n)$ of invertible linear transformations of the space K^n. Accordingly, every matrix representation of a group G can be regarded as a group homomorphism of G into $\mathrm{GL}(K^n)$.

Example. Let M be the matrix representation of the group S_n constructed above in 0.2. For each $\sigma \in S_n$ the matrix $M(\sigma)$ can be regarded as a linear transformation of K^n. Let $(e_1, \dots, e_n)$ be the standard basis of K^n. Then

$$(7) \qquad M(\sigma)e_i = e_{\sigma(i)} \qquad \text{for} \quad i = 1, \dots, n.$$

Since a linear operator is uniquely determined by its action on the basis vectors, equations (7) may be taken as the definition of the representation M.

Replacing K^n by an arbitrary vector space V over the field K, we now arrive at the following generalization of the notion of a matrix representation.

Definition. A LINEAR REPRESENTATION of the group G over the field K is a homomorphism of G into the group $\mathrm{GL}(V)$ of all invertible linear transformations (linear operators) of a vector space V over K. V is called the REPRESENTATION SPACE, and its dimension is called the DIMENSION or the DEGREE of the representation.

Suppose that the space V has a finite dimension n. Then with each linear representation T of the group G in V we can associate a class of n-dimensional matrix representations. To this end we pick some basis $(e) = (e_1, \dots, e_n)$ of V. Every operator $T(g)$, for $g \in G$, is described with respect to the basis (e) by a matrix $T(g)_{(e)}$, and the map $T_{(e)} : g \mapsto T(g)_{(e)}$ that arises in this manner is a matrix representation of G. On choosing another basis $(f) = (e)C$ (where C is the transition matrix from (e) to (f)), we obtain another representation $T_{(f)}$, related to $T_{(e)}$ as follows:

$$T_{(f)}(g) = C^{-1}T_{(e)}(g)C.$$

Definition. We say that two matrix representations T_1 and T_2 are EQUIVALENT (and write $T_1 \simeq T_2$) if they have the same dimension and there exists a nonsingular matrix C such that

$$T_2(g) = C^{-1}T_1(g)C \tag{8}$$

for all $g \in G$.

The foregoing discussion makes clear that to each finite-dimensional representation there corresponds a class of equivalent matrix representations.

0.5. Definition. We say that two linear representations,

$$T_1 : G \to \mathrm{GL}(V_1) \qquad \text{and} \qquad T_2 : G \to \mathrm{GL}(V_2),$$

are ISOMORPHIC (or EQUIVALENT), and write $T_1 \simeq T_2$, if there exists an isomorphism $\sigma : V_1 \to V_2$ such that

$$\sigma T_1(g) = T_2(g)\sigma \tag{9}$$

for all $g \in G$.

In other words, $T_1 \simeq T_2$, if upon identifying the spaces V_1 and V_2 by means of σ, the representations T_1 and T_2 become identical. In particular, if V_1 and V_2 are finite-dimensional, then in bases that correspond under σ the matrices of the operators $T_1(g)$ and $T_2(g)$ coincide for all $g \in G$. This means that the matrix representations associated with the linear representations T_1 and T_2 are identical for a compatible choice of bases, and equivalent for an arbitrary choice of bases.

Conversely, suppose that for some choice of bases of the spaces V_1 and V_2 the linear representations T_1 and T_2 determine equivalent matrix representations. Then, for an appropriate choice of bases, T_1 and T_2 determine identical matrix representations and hence are isomorphic.

Specifically, let the bases $(e)_1$ and and $(e)_2$ in V_1 and V_2 be such that $T_1(g)_{(e)_1} = T_2(g)_{(e)_2}$ for all $g \in G$. Then the isomorphism $\sigma : V_1 \to V_2$ which takes $(e)_1$ into $(e)_2$ satisfies condition (9).

Thus, *the matrix representations associated with the finite-dimensional linear representations T_1 and T_2 are equivalent if and only if T_1 and T_2 are isomorphic.*

0.6. Halting here this somewhat unexciting yet necessary discussion, let us see what benefits we can extract from it.

Recall that linear operators can be added, multiplied with one another, and multiplied by numbers (elements of the ground field).

In an arbitrary basis, to these operations there correspond the same operations on matrices; however, the definitions of the operations on linear operators are independent of the choice of a basis. Further, in a finite-dimensional space over $\mathbf{R}$ or $\mathbf{C}$, one can define the limit of a sequence of linear operators. The act of passing to the limit is also compatible with the corresponding action on matrices, but again does not depend on the choice of a basis.

Let V be a finite-dimensional vector space over $K = \mathbf{R}$ or $\mathbf{C}$. The exponential operator-valued function

$$t \mapsto \mathrm{e}^{t\alpha} \qquad (t \in K, \quad \alpha \in \mathrm{L}(V))$$

can be defined in the same manner as the matrix exponential function (see 0.1 above). It is a linear representation of the additive group K. The matrix representation that corresponds to this representation in an arbitrary basis is $t \mapsto \mathrm{e}^{tA}$, where A denotes the matrix of the operator α in that basis. Since the choice of basis is at our disposal, we can attempt to make it so that A will take the simplest possible form. For example, in the case $K = \mathbf{C}$ we can arrange that A be in Jordan form. It is known that the Jordan form is uniquely determined up to a permutation of the blocks. This implies, in particular, that two linear representations,

$$t \mapsto \mathrm{e}^{t\alpha} \qquad \text{and} \qquad t \mapsto \mathrm{e}^{t\beta},$$

are isomorphic if and only if the matrices of the operators α and β have the same Jordan form.

This example illustrates the geometric approach to the notion of representation, which does not distinguish between equivalent matrix representations and permits us, in certain cases, to avoid computations in coordinates.

0.7. In order to describe a linear representation, it is not obligatory to choose a basis in the representation space. Alternatively, representations can be described geometrically.

Examples.

1. We specify a linear representation of the group $\mathbf{R}$ as follows: to the element $t \in \mathbf{R}$ we assign the rotation by angle t in the Euclidean plane. From geometric considerations it is obvious that the composition of the rotations by the angles t and u is the rotation by the angle $t + u$, and hence that the map thus constructed is indeed a linear representation. In an orthonormal basis we have the corresponding matrix representation (4). Formula (3) now follows automatically, and we can derive the addition formulas for trigonometric functions from it (and not conversely, as we did in 0.1).

2. Let V denote the linear space of all polynomials with real coefficients. To each $t \in \mathbf{R}$ we assign a linear operator $L(t) \in \mathrm{GL}(V)$ by the rule

$$(L(t)f)(x) = f(x-t). \tag{10}$$

It is readily checked that $L(t)$ is indeed a linear operator and that $L(t+u) = L(t)L(u)$, i.e., L is a linear representation of the additive group $\mathbf{R}$.

3. In the preceding example we can replace polynomials by any space of functions which is invariant under translations. Here are some examples:

a) the space of continuous functions;

b) the space of polynomials of degree $\leq n$;

c) the space of trigonometric polynomials, i.e., polynomials in $\cos x$ and $\sin x$;

d) the linear span of the functions $\cos x$ and $\sin x$.

Let us examine in more detail the last case. Since

$$\begin{aligned} L(t)\cos x &= \cos t \cos x + \sin t \sin x, \\ L(t)\sin x &= -\sin t \cos x + \cos t \sin x, \end{aligned} \tag{11}$$

the transformation $L(t)$ takes any linear combination of $\cos x$ and $\sin x$ again into such a linear combination. Formulas (11) show that in the basis $(\cos x, \sin x)$ the operator $L(t)$ is described by the matrix

$$\begin{pmatrix} \cos t & -\sin t \\ \sin t & \cos t \end{pmatrix}.$$

Therefore, the representation L of the group $\mathbf{R}$ in the space $\langle \cos x, \sin x \rangle$ is isomorphic to the representation constructed above geometrically.

0.8. Group theory is also concerned with more general "representations," called *actions*.

Let X be an arbitrary set and let $S(X)$ denote the group of all bijections of X onto itself. (If $X = \{1, 2, \ldots, n\}$, then $S(X) = S_n$).

Definition. An ACTION of the group G on X is a homomorphism $s\colon G \to S(X)$.

In other words, the action s assigns to each $g \in G$ a bijective map $s(g)$ of the set X onto itself in such a manner that $s(g_1 g_2) = s(g_1)s(g_2)$ for all $g_1, g_2 \in G$.

If it is clear to which action we are referring, we will simply write gx instead of $s(g)x$. The last equality above signified "associativity": $(g_1g_2)x = g_1(g_2x)$ for every $x \in X$.

We may regard a linear representation as a special kind of action. As a nonlinear example we give the action of $\mathbf{R}$ on itself defined by the rule $s(t)x = x + t$.

This last example can be generalized as follows. Let G be a group. Then G acts on itself by left translations,

(12) $\qquad l(g)x = gx,$

as well as by right translations,

(13) $\qquad r(g)x = xg^{-1}.$

Let us check that, say, formula (13) determines an action:

$$r(g_1g_2)x = x(g_1g_2)^{-1} = xg_2^{-1}g_1^{-1},$$

and

$$r(g_1)r(g_2)x = r(g_1)(xg_2^{-1}) = xg_2^{-1}g_1^{-1}.$$

(The proof explains why g appears in (13) with the exponent -1.)

0.9. With each action we can associate a *linear representation in a function space.*

Let K be a field and $K[X]$ the vector space of all K-valued functions on the set X. Each $\sigma \in S(X)$ defines a linear transformation σ_* in $K[X]$:

(14) $\qquad (\sigma_* f)(x) = f(\sigma^{-1}x) \qquad (f \in K[X]).$

We have $(\sigma\tau)_* = \sigma_*\tau_*$, i.e., (14) defines a linear representation of the group $S(X)$. In fact,

$$((\sigma\tau)_* f)(x) = f((\sigma\tau)^{-1}x) = f(\tau^{-1}\sigma^{-1}x),$$

and

$$(\sigma_*\tau_* f)(x) = (\tau_* f)(\sigma^{-1}x) = f(\tau^{-1}\sigma^{-1}x).$$

In an analogous manner we can define a linear representation of the group $S(X)$ in a space of functions of several variables:

$$(\sigma_* f)(x_1, \ldots, x_k) = f(\sigma^{-1}x_1, \ldots, \sigma^{-1}x_k).$$

Now suppose we are given an action $s\colon G \to S(X)$. We define a linear representation s_* of the group G in the space $K[X]$, setting

$$s_*(g) = s(g)_*.$$

Representations of G in spaces of functions of several variables are defined similarly.

Examples.

1. Consider the group G of rotations of a cube (isomorphic, as is known, to S_4) and its natural action s on the set X of faces of the cube. Here the space $K[X]$ is six-dimensional. As a basis of $K[X]$ we can take the functions $f_1, \dots, f_6$, each of which is equal to 1 on one of the faces and to 0 on the others. Relative to this basis the operators $s_*(g)$ are written as matrices of 0's and 1's such that in every row and every column there is exactly one 1. For example, let g be the rotation by $2\pi/3$ around an axis passing through the center of the cube and one of its vertices. Then for a suitable labeling of the basis functions $f_1, \dots, f_6$, the operator $s_*(g)$ is given by the matrix

$$\begin{pmatrix} 0 & 0 & 1 & 0 & 0 & 0 \\ 1 & 0 & 0 & 0 & 0 & 0 \\ 0 & 1 & 0 & 0 & 0 & 0 \\ 0 & 0 & 0 & 0 & 0 & 1 \\ 0 & 0 & 0 & 1 & 0 & 0 \\ 0 & 0 & 0 & 0 & 1 & 0 \end{pmatrix}$$

2. Let s be the natural action of the group S_n on the set $\{1, 2, \dots, n\}$. Then the representation s_* is isomorphic to the representation M constructed above in 0.2 (see also the example in 0.4).

In general, if X is a finite set, then the space $K[X]$ is finite-dimensional and its dimension is equal to the number of elements of X. The functions δ_x, $x \in X$, defined as

$$\delta_x(y) = \begin{cases} 1 & \text{if } y = x, \\ 0 & \text{if } y \neq x, \end{cases} \tag{15}$$

form a basis of $K[X]$. The operators $s_*(g)$, $g \in G$, simply permute the elements of this basis. Specifically,

$$s_*(g)\delta_x = \delta_{gx}. \tag{16}$$

In fact,

$$(s_*(g)\delta_x)(y) = \delta_x(g^{-1}y) = \begin{cases} 1 & \text{if } y = gx, \\ 0 & \text{if } y \neq gx. \end{cases}$$

In those cases of interest where the set X is infinite, it usually possesses an additional structure (for example, it is a topological space or a differentiable manifold), and the given group action is compatible with that structure. Then one considers only functions that are "nice" in one sense or another (for example, continuous or differentiable), rather than arbitrary functions on X. For instance, in the case of the action s of the group $\mathbf{R}$ on itself by translations, we can take the space of polynomials for the representation space of s_*. We thus get the linear representation of $\mathbf{R}$ considered above in Example 2, 0.7.

Definition. The linear representation $l_* = L$ of an arbitrary group G (in a suitable class of functions on G) associated with the action l of G on itself by left translations is called the LEFT REGULAR REPRESENTATION of G.

According to this definition

$$(17) \qquad (L(g)f)(x) = f(g^{-1}x) \qquad (g, x \in G).$$

The RIGHT REGULAR REPRESENTATION is defined in a similar manner. Needless to say, the class of functions in question must be specified exactly.

As we shall see, the study of regular representations is the key step towards the description of all representations of a given group.

0.10. Methods of producing new representations from one or several given ones play an important role in the theory of linear representations. One of the simplest of these methods is *the composition of a representation and a homomorphism*.

Let $T: G \to \mathrm{GL}(V)$ be a linear representation of the group G, and let $\phi: H \to G$ be a homomorphism. Then $T \circ \phi$ is a linear representation of the group H.

Let us examine two particular cases of this construction. If H is a subgroup of G and ϕ is the inclusion map of H into G, then $T \circ \phi$ is simply *the restriction of the representation T to H*. We denote the restriction operation by Res_H^G. According to the definition,

$$(\mathrm{Res}_H^G T)(h) = T(h) \qquad \text{for } h \in H.$$

If now ϕ is an *automorphism* of G, then $T \circ \phi$ is, like T, a representation of G. Such a "twisted" representation can be isomorphic or not to the original representation T.

Examples.

1. Let $\phi = a(h)$ be the *inner automorphism* defined by $h \in G$, i.e., $a(h)g = hgh^{-1}$. Then for every $g \in G$

$$(T \circ a(h))(g) = T(hgh^{-1}) = T(h)T(g)T(h)^{-1}$$

or, equivalently,

$$(18) \qquad (T \circ a(h))(g)T(h) = T(h)T(g).$$

Since $T(h)$ is an isomorphism of the vector space V onto itself, equality (18) shows that $T \circ a(H) \simeq T$.

2. Let ϕ be the automorphism of the group $\mathbf{C}$ acting as $\phi(x) = -x$. Let V be a finite-dimensional complex vector space. The map $F_\alpha: t \mapsto \mathrm{e}^{t\alpha}$ is a linear

representation of $\mathbf{C}$ for every $\alpha \in \mathrm{L}(V)$ (see 0.6). Obviously, $F_\alpha \circ \phi = F_{-\alpha}$. The representations F_α and $F_{-\alpha}$ are isomorphic if and only if the matrices of the operators α and $-\alpha$ have the same Jordan form. (The latter in turn holds if and only if for any k and c in the Jordan form of the matrix of α, the number of Jordan blocks of order k with eigenvalue c is equal to the number of Jordan blocks of order k with eigenvalue $-c$).

Questions and Exercises

1.* Show that $\det \mathrm{e}^A = \mathrm{e}^{\operatorname{tr} A}$ for any matrix $A \in \mathrm{L}_n(\mathbf{R})$.

2. If F is as given below, show that F is a matrix representation of $\mathbf{R}$ and find a matrix A such that $F(t) = \mathrm{e}^{tA}$:

a) $$F(t) = \begin{pmatrix} \cosh t & -\sinh t \\ \sinh t & \cosh t \end{pmatrix};$$

b) $$F(t) = \begin{pmatrix} 1 & t \\ 0 & 1 \end{pmatrix}.$$

3. What is a one-dimensional matrix representation?

4. Let M be the matrix representation of the group S_n constructed in 0.2. Show that $\operatorname{tr} M(\sigma)$ is the number of fixed points of the permutation σ.

5. How many trivial matrix representations does an arbitrary group admit?

6. Show, without resorting to calculations, that $\mathrm{e}^{C^{-1}AC} = C^{-1}\mathrm{e}^A C$ for any matrices $A \in \mathrm{L}_n(\mathbf{R})$ and $C \in \mathrm{GL}_n(\mathbf{R})$.

7. Let $S\colon \mathbf{R} \to S(V)$ be one of the maps listed below, where V is the space of all polynomials with real coefficients and $t \in \mathbf{R}, f \in V$:

a) $(S(t)f)(x) = f(tx)$;

b) $(S(t)f)(x) = f(\mathrm{e}^t x)$;

c) $(S(t)f)(x) = \mathrm{e}^t f(x)$;

d) $(S(t)f)(x) = f(x) + t$;

e) $(S(t)f)(x) = \mathrm{e}^t f(x + t)$.

Is S a linear representation?

8. Describe one of the equivalent matrix representations associated with the linear representation $S\colon \mathbf{R} \to \mathrm{GL}(V)$, where V is the space of polynomials of degree ≤ 3 and $(S(t)f)(x) = f(x-t)$ for $t \in \mathbf{R}, f \in V$.

9. Find all finite-dimensional linear representations of

a) $\mathbf{Z}$;

b) $\mathbf{Z}_m$.

10.* Find all differentiable finite-dimensional complex linear representations of

a) $G = \mathbf{R}^+$ (the multiplicative group of positive reals);

b) $G = \mathbf{T} = \{ z \in \mathbf{C}^* \mid |z| = 1 \}$.

11. Let s be an action of the group G on a set X and e the identity element of G. Show that $s(e)$ is the identity map of X.

12. Let $\hat{\mathbf{R}} = \mathbf{R} \cup \{\infty\}$. For any matrix $A = \begin{pmatrix} a & b \\ c & d \end{pmatrix} \in \mathrm{GL}_2(\mathbf{R})$ put

$$s(A)x = \frac{ax+b}{cx+d} \qquad (x \in \hat{\mathbf{R}}),$$

with the convention that

$$\frac{a \cdot \infty + b}{c \cdot \infty + d} = \frac{a}{c} \quad \text{and} \quad \frac{u}{0} = \infty$$

for $u \neq 0$. Show that s is an action of $\mathrm{GL}_2(\mathbf{R})$ on $\hat{\mathbf{R}}$.

13. Write down an explicit formula for the right regular representation of the group G.

14.* Prove that the left regular representation of any group G is isomorphic to its right regular representation.

15. Show that every group admits a faithful linear representation.

16. Is every finite-dimensional complex representation of $\mathbf{Z}$ obtained by restricting a representation of $\mathbf{C}$ to $\mathbf{Z}$?

17. Let ϕ denote the automorphism of the group $\mathbf{Z}_m$ defined by the rule $\phi(x) = -x$. Find all complex finite-dimensional representations T of $\mathbf{Z}_m$ with the property that $T \circ \phi \simeq T$.

I. General Properties of Representations

In this chapter we give an account of the basic definitions and simplest theorems of the theory of linear representations. Some of these theorems are valid for both finite- and infinite-dimensional representations. This textbook, however, is devoted to the former case representations, and the reader is practically at no loss if he assumes that all representations considered here are *finite-dimensional* (except, of course, for those examples which are manifestly infinite-dimensional).

1. Invariant Subspaces

1.1. The study of the structure of linear representations begins with that of invariant subspaces.

Definition. Let $T: G \to \mathrm{GL}(V)$ be a linear representation of the group G in a vector space V. A subspace $U \subset V$ is said to be INVARIANT UNDER REPRESENTATION T (or G-INVARIANT, if it is clear which representation of G one has in mind) if

(1) $\qquad T(g)u \in U \qquad$ for all $g \in G$ and $u \in U$.

For example, let L be the representation of the additive group $\mathbf{R}$ in the space of all polynomials, given by the rule

$$(L(t)f)(x) = f(x-t).$$

Then the subspace of all polynomials of degree $\leq n$ is invariant under L for every n.

It is obvious that *sums and intersections of invariant subspaces are invariant.*

Suppose that the space V is finite-dimensional and that $(e) = (e_1, \dots, e_n)$ is some basis in V such that $U = \langle e_1, \dots, e_k \rangle$. Then the invariance of U under

E.B. Vinberg, *Linear Representations of Groups*, Modern Birkhäuser Classics,
DOI 10.1007/978-3-0348-0063-1_2,

the given representation T of G means that in the basis (e) each operator $T(g)$, $g \in G$, is given by a matrix of the form

$$(2) \qquad T_{(e)}(g) = \left(\begin{array}{c|c} A(g) & C(g) \\ \hline \underbrace{\quad 0 \quad}_{k} & B(g) \end{array}\right) \begin{matrix} \}\, k \\ \end{matrix} .$$

1.2. With every invariant subspace U we can associate two linear representations of G, acting on the spaces U and V/U respectively. The first of these, called a SUBREPRESENTATION of T and denoted by T_U, is obtained by restricting the operators $T(g)$ to U:

$$(3) \qquad T_U(g) = T(g)\big|_U \qquad \text{for all } g \in G.$$

The second, called a QUOTIENT or FACTOR REPRESENTATION of T and denoted by $T_{V/U}$, is defined as follows:

$$(4) \qquad T_{V/U}(x + U) = T(g)x + U \qquad \text{for all } g \in G, \quad x \in V.$$

(Recall that the elements of the quotient space V/U are the cosets $x + U$ with $x \in V$.)

Definition (4) requires some further explanations. First of all, we have to verify that the right-hand side does not depend upon the choice of the representative x in a given coset. Replacing x by $x' = x + u$ with $u \in U$, we get

$$T(g)x' + U = T(g)x + T(g)u + U = T(g)x + U.$$

Here we have used in an essential manner the invariance of U, which guarantees that $T(g)u \in U$. Next, we have to check that $T_{V/U}(g)$ is a linear operator. By the addition rule for cosets,

$$\begin{aligned} T_{V/U}(g)((x + U) + (y + U)) &= T_{V/U}(g)(x + y + U) \\ &= T(g)(x + y) + U \\ &= T(g)x + T(g)y + U \\ &= (T(g)x + U) + (T(g)y + U) \\ &= T_{V/U}(g)(x + U) + T_{V/U}(g)(y + U). \end{aligned}$$

The homogeneity of $T_{V/U}$ is verified in a similar manner. Finally, we have to show that the map $g \mapsto T_{V/U}(g)$ is a homomorphism, i.e.,

$$T_{V/U}(g_1 g_2) = T_{V/U}(g_1) T_{V/U}(g_2) \qquad \text{for all } g_1, g_2 \in G.$$

But this is a straightforward consequence of the definition of $T_{V/U}$ and the equality $T(g_1 g_2) = T(g_1) T(g_2)$.

If the space V is finite-dimensional, T_U and $T_{V/U}$ can be conveniently described in terms of matrices. To this end, we pick a basis $(e_1, \ldots, e_n)$ of V

such that $U = \langle e_1, \dots, e_k \rangle$. Then the operators $T(g)$, $g \in G$, are described by the matrices (2). Here $A(g)$ and $B(g)$ are the matrices of the operators $T_U(g)$ and $T_{V/U}(g)$ in the bases $(e_1, \dots, e_k)$ of U and $(e_{k+1} + U, \dots, e_n + U)$ of V/U respectively.

To prove the assertion concerning the matrix $B(g)$, let $b_{ij}(g)$ (respectively $c_{ij}(g)$) denote the entry of $B(g)$ (respectively $C(g)$) lying on the i-th row and j-th column of the matrix $T_{(e)}(g)$. Then for every $j > k$

$$\begin{aligned} T_{V/U}(g)(e_j + U) &= T(g)e_j + U \\ &= \sum_{i=1}^{k} c_{ij}(g)e_i + \sum_{i=k+1}^{n} b_{ij}(g)e_i + U \\ &= \sum_{i=k+1}^{n} b_{ij}(g)e_i + U = \sum_{i=k+1}^{n} b_{ij}(g)(e_i + U), \end{aligned}$$

as it should be.

1.3. Definition. A linear representation $T: G \to \mathrm{GL}(V)$ is said to be IRREDUCIBLE if there are no nontrivial (i.e., different from 0 and V) subspaces $U \subset V$ invariant under T.

Examples.

1. Every one-dimensional representation is irreducible.

2. The identity representation of $\mathrm{GL}(V)$ is irreducible, since every nonnull vector in V can be taken into any other such vector by an invertible linear transformation.

3. The representation of $\mathbf{R}$ by rotations in the plane (see Example 1, 0.7) is also irreducible.

4. The representation of $\mathbf{R}$ by translations in the space of polynomials (see Example 2, 0.7) is not irreducible.

5. Let V be an n-dimensional vector space over the field K, and let $e_1, \dots, e_n$ be a basis in V. The representation M of the group S_n in V specified by the rule

$$M(\sigma)e_i = e_{\sigma(i)} \qquad (i = 1, \dots, n)$$

(see 0.2 and 0.4) is called a MONOMIAL REPRESENTATION OF S_n. It is not irreducible: for example, it leaves invariant the $(n-1)$-dimensional subspace

$$V_0 = \left\{ \sum x_i e_i \,\middle|\, \sum x_i = 0 \right\},$$

and also the one-dimensional subspace

$$V_1 = \left\langle \sum e_i \right\rangle.$$

If the characteristic of the field K is equal to zero, then $V_1 \not\subset V_0$, and hence

$$V = V_0 \oplus V_1.$$

We claim that in this case the representation $M_0 = M_{V_0}$ is irreducible. In fact, suppose $U \subset V_0$ is an invariant subspace. Let $x = \sum x_i e_i$ be a nonnull vector in U. Since $x \notin V_1$, at least two of the numbers x_i are distinct. Suppose, for the sake of definiteness, that $x_1 \neq x_2$. Then

$$M((12))x - x = (x_2 - x_1)(e_1 - e_2) \in U,$$

whence $e_1 - e_2 \in U$. Applying to $e_1 - e_2$ various operators $M(\sigma)$, we can obtain all vectors of the form $e_i - e_j$, and the latter span the subspace V_0. Thus $U = V_0$, as we needed to show.

1.4. Definition. The linear representation $T : G \to \mathrm{GL}(V)$ is said to be COMPLETELY REDUCIBLE if every invariant subspace $U \subset V$ has an invariant complement W. (Recall that W is called a COMPLEMENT of U if $V = U \oplus W$.)

Every irreducible representation is completely reducible (though from the point of view of the Russian [or English] language this may sound strange!) In fact, for an irreducible representation there are only two invariant subspaces: the entire representation space and the null subspace, which complement one another. Hence every invariant subspace has an invariant complement.

Notice that if U and W are complementary subspaces, then the restriction σ of the canonical map $V \to V/U$ to W is an isomorphism of the space W onto V/U (each coset of U in V contains exactly one element from W). If, in addition, U and W are invariant under the representation $T: G \to \mathrm{GL}(V)$, then σ commutes with the action of G:

$$\sigma T_W(g)x = T(g)x + U = T_{V/U}(g)\sigma x.$$

This implies that *the representations T_W and $T_{V/U}$ are isomorphic.*

Let us examine in more detail the finite-dimensional case. Let $T: G \to \mathrm{GL}(V)$ be a finite-dimensional linear representation of the group G. Let $U, W \subset V$ be complementary invariant subspaces. Pick bases $(e_1, \dots, e_k)$ in U and $(e_{k+1}, \dots, e_n)$ in W. Together they yield a basis $(e) = (e_1, \dots, e_n)$ in V. Relative to (e), the operators $T(g)$, for $g \in G$, are given by matrices of the form

$$\left(\begin{array}{c|c} A(g) & 0 \\ \hline \underbrace{0}_{k} & B(g) \end{array}\right) \Big\}\, k, \tag{5}$$

where $A(g)$ is the matrix of $T_U(g)$ in the basis $(e_1, \dots, e_k)$, and $B(g)$ is the matrix of $T_W(g)$ in the basis $(e_{k+1}, \dots, e_n)$, as well as the matrix of $T_{V/U}(g)$ (see 1.2 above).

Example. Consider the two-dimensional representations F and S of $\mathbf{R}$, given in the basis $(e) = (e_1, e_2)$ by the matrices

$$F_{(e)}(t) = \begin{pmatrix} 1 & t \\ 0 & 1 \end{pmatrix} \quad \text{and} \quad S_{(e)}(t) = \begin{pmatrix} \mathrm{e}^t & \mathrm{e}^t - 1 \\ 0 & 1 \end{pmatrix}.$$

In both cases $U = \langle e_1 \rangle$ is an invariant subspace. Does it admit an invariant complement?

In the first case, F_U and $F_{V/U}$ are trivial representations. Assuming that an invariant complement to U exists, F would be specified, in a suitable basis, by the (t-independent) matrix

$$\begin{pmatrix} 1 & 0 \\ 0 & 1 \end{pmatrix},$$

i.e., would be a trivial representation, which is not the case. Thus U has no invariant complement. In particular, F is not completely reducible.

For representation S, one can check that $S(t)(e_1 - e_2) = e_1 - e_2$ for all $t \in \mathbf{R}$. Consequently, $\langle e_1 - e_2 \rangle$ is an invariant subspace. Relative to the basis $(e_1, e_1 - e_2)$, S is given by the diagonal matrix

$$\begin{pmatrix} \mathrm{e}^t & 0 \\ 0 & 1 \end{pmatrix}.$$

It is readily verified that $\langle e_1 \rangle$ and $\langle e_1 - e_2 \rangle$ are the only nontrivial subspaces invariant under S. This shows that S is completely reducible.

1.5. Theorem 1. *Every subrepresentation of a completely reducible representation is completely reducible.*

PROOF. Let $T : G \to \mathrm{GL}(V)$ be a completely reducible representation, $U \subset V$ an invariant subspace, and U_1 an arbitrary invariant subspace contained in U. Since T is completely reducible, U_1 has an invariant complement W in V. Consider the subspace $W \cap U$. It is invariant and, as is readily verified, $U = U_1 \oplus (W \cap U)$, i.e., $W \cap U$ is a complement of U_1 in U. This proves the complete reducibility of the representation T_U. $\square$

Theorem 2. *The representation space of any completely reducible finite-dimensional representation admits a decomposition into a direct sum of minimal invariant subspaces.*

(We call an invariant subspace MINIMAL if it is minimal among the nonzero invariant subspaces.)

PROOF. We proceed by induction on the dimension of the representation space. Let $T : G \to \mathrm{GL}(V)$ be a completely reducible representation. If T is irreducible, the theorem is plainly true (and the sum reduces to one term). In the opposite case there exist nontrivial invariant subspaces. Let U be an arbitrary minimal invariant subspace, and let W be an invariant complement of U. By Theorem 1, the representation T_W is completely reducible. Applying the inductive hypothesis to T_W, we can assume that W decomposes into a direct sum of minimal invariant subspaces. Adding U to this decomposition, we obtain a decomposition of V into a direct sum of minimal invariant subspaces. □

Theorem 3. *Let $T : G \to \mathrm{GL}(V)$ be a linear representation. Let*

$$V = V_1 + V_2 + \ldots + V_m \tag{6}$$

be a decomposition of the space V into a (not necessarily direct) sum of minimal invariant subspaces. Then T is completely reducible. Moreover, for every invariant subspace U there exist indices $i_1, \ldots, i_p$ such that

$$V = U \oplus V_{i_1} \oplus \ldots \oplus V_{i_p}. \tag{7}$$

PROOF. It suffices to prove the second assertion of the theorem, since it is a stronger version of the first. Let U be an invariant subspace, and let $\{i_1, \ldots, i_p\}$ be a (possibly empty) maximal set of indices such that the subspaces $U, V_{i_1}, \ldots, V_{i_p}$ are linearly independent. We claim that (7) holds in this case. It suffices to show that

$$V_i \subset U \oplus V_{i_1} \oplus \ldots \oplus V_{i_p} \tag{8}$$

for every $i \notin \{i_1, \ldots, i_p\}$. Since $V_i \cap (U \oplus V_{i_1} \oplus \ldots \oplus V_{i_p})$ is an invariant subspace contained in V_i, and since V_i is a minimal invariant subspace, either (8) holds or

$$V_i \cap (U \oplus V_{i_1} \oplus \ldots \oplus V_{i_p}) = 0. \tag{9}$$

However, alternative (9) is impossible, because it would imply the linear independence of the subspaces $U, V_{i_1}, \ldots, V_{i_p}$, and V_i, thereby contradicting the choice of the set $\{i_1, \ldots, i_p\}$. This completes the proof of the theorem. □

Remarks.

1) Applying Theorem 3 to the subspace $U = 0$ we conclude that V itself is the direct sum of a number of subspaces V_i.

2) An invariant subspace is not necessarily the direct sum of a number of subspaces V_i. For instance, let T be the trivial representation in V. Then every subspace of V is invariant, and the minimal invariant subspaces are precisely the one-dimensional ones. Let $(e_1, \ldots, e_n)$ be an arbitrary basis in V. Then $V = \langle e_1 \rangle \oplus \ldots \oplus \langle e_n \rangle$ is a decomposition of V into a direct sum of minimal invariant subspaces. However, if $n > 1$, not every subspace is the linear span of a subset of basis vectors.

1.6. Some Examples. We consider three linear representations of the group $\mathrm{GL}(V)$, where V is an n-dimensional vector space over K.

1. *Representation by left multiplication in the algebra* $\mathrm{L}(V)$ *of all linear operators in* V:

$$\Lambda(\alpha)\xi = \alpha\xi \qquad (\alpha \in \mathrm{GL}(V), \quad \xi \in \mathrm{L}(V)).$$

Linear operators can be replaced by their matrices in some fixed basis of V. Then the definition of the representation Λ is accordingly modified to

$$\Lambda(A)X = AX \qquad (A \in \mathrm{GL}_n(K), \quad X \in \mathrm{L}_n(K)). \tag{10}$$

Let $\mathrm{L}^{(i)}$ denote the subspace of all matrices for which every column except the i-th contains only zeros. Obviously,

$$\mathrm{L}_n(K) = \mathrm{L}^{(1)} \oplus \mathrm{L}^{(2)} \oplus \ldots \oplus \mathrm{L}^{(n)}. \tag{11}$$

On multiplying the matrix X on the left by the matrix A, every column of X is multiplied by A. This implies that the subspaces $\mathrm{L}^{(i)}$ are invariant under Λ. Moreover, $\Lambda_{\mathrm{L}^{(i)}}$ is isomorphic to the identity representation Id of $\mathrm{GL}_n(K)$ in the space of columns, and hence it is irreducible. According to Theorem 3, the representation Λ is completely reducible.

To the decomposition (11) there corresponds a decomposition of the space $\mathrm{L}(V)$ into a direct sum of minimal invariant subspaces. Such a decomposition is not unique: changing the basis in V, generally speaking, also changes the decomposition.

2. *The* ADJOINT REPRESENTATION *in the algebra* $\mathrm{L}(V)$:

$$\mathrm{Ad}(\alpha)\xi = \alpha\xi\alpha^{-1} \qquad (\alpha \in \mathrm{GL}(V), \quad \xi \in \mathrm{L}(V)).$$

This is indeed a representation:

$$\mathrm{Ad}(\alpha\beta)\xi = \alpha\beta\xi(\alpha\beta)^{-1} = \alpha\beta\xi\beta^{-1}\alpha^{-1} = \mathrm{Ad}(\alpha)\mathrm{Ad}(\beta)\xi.$$

In terms of matrices Ad is defined as follows:

$$\text{(12)} \qquad \mathrm{Ad}(A)X = AXA^{-1} \qquad (A \in \mathrm{GL}_n(K), \quad X \in \mathrm{L}_n(K)).$$

One can show that there are only two nontrivial Ad-invariant subspaces: the one-dimensional subspace $\langle E\rangle$ and the (n^2-1)-dimensional subspace $\mathrm{L}_n^0(K)$ of the matrices with trace zero. If the characteristic of K is equal to zero, then $E \notin \mathrm{L}_n^0(K)$, and so

$$\mathrm{L}_n(K) = \langle E\rangle \oplus \mathrm{L}_n^0(K).$$

In this case the adjoint representation is completely reducible.

3. *The representation in the space* $\mathrm{B}(V)$ *of bilinear functions (forms) on* V:

$$(\Phi(\alpha)f)(x,y) = f(\alpha^{-1}x, \alpha^{-1}y) \qquad (\alpha \in \mathrm{GL}(V), \quad f \in \mathrm{B}(V)).$$

This is a natural definition if one is guided by the general principle that every one-to-one mapping σ of an arbitrary set X onto itself acts on functions of one or several X-valued arguments if one applies σ^{-1} simultaneously to all arguments (see 0.9). It is readily verified that $\Phi(\alpha)f$ is again a bilinear function.

The subspaces $\mathrm{B}^+(V)$ and $\mathrm{B}^-(V)$ of symmetric and skew-symmetric bilinear functions are invariant under Φ. If the characteristic of K is different from two, then

$$\text{(13)} \qquad \mathrm{B}(V) = \mathrm{B}^+(V) \oplus \mathrm{B}^-(V),$$

and one can show that $\mathrm{B}^+(V)$ and $\mathrm{B}^-(V)$ are minimal. In this case Φ is completely reducible.

In terms of matrices, Φ is defined as follows:

$$\text{(14)} \qquad \Phi(A)X = (A^{-1})'XA^{-1} \qquad (A \in \mathrm{GL}_n(K), \quad X \in \mathrm{L}_n(K)),$$

where $'$ stands for transposition of matrices.

To (13) there corresponds the decomposition

$$\mathrm{L}_n(K) = \mathrm{L}_n^+(K) \oplus \mathrm{L}_n^-(K),$$

where $\mathrm{L}_n^+(K)$ and $\mathrm{L}_n^-(K)$ denote the spaces of symmetric and skew-symmetric matrices respectively.

In the following we shall, as a rule, write $\alpha_* f$ instead of $\Phi(\alpha)f$, in agreement with the general notation adopted in 0.9.

Questions and Exercises

1. Prove that if the subspace U of the space of the representation T of G is invariant, then $T(g)U = U$ for all $g \in G$.

2. Find all subspaces of the space of polynomials that are invariant under the representation L of $\mathbf{R}$ given by the formula $(L(t)f)(x) = f(x-t)$.

3. Let F denote the representation of $\mathbf{C}$ in a complex n-dimensional space given by the formula $F(t) = e^{t\alpha}$, where α is a linear operator whose characteristic polynomial has no multiple roots. Find all subspaces invariant under F.

4. Without resorting to computations, prove that the matrix $B(g)$ in formula (2) does not change on passing to a new basis $(f) = (f_1, \dots, f_n)$ of the space V if $f_i - e_i \in U$ for all $i > k$.

5. Let W_1 and W_2 be two invariant complements of the invariant subspace U in the space of the representation T. Prove that $T_{W_1} \simeq T_{W_2}$.

6. Prove that every quotient representation of a completely reducible representation is completely reducible.

7. Is the representation

a) of Exercise 2,

b) of Exercise 3

completely reducible?

8. Prove that the representation $t \mapsto e^{t\alpha}$ of $\mathbf{C}$ is completely reducible if and only if the operator α is diagonalizable (i.e., it admits a basis of eigenvectors).

9. Prove that the identity representation of the orthogonal group O_n is irreducible for any n.

10. Prove that any monomial representation of the group S_n over a field of characteristic zero is completely reducible.

11. Prove that for $n \geq 4$ the restriction of the representation M_0 (see Example 5 of 1.3) to the subgroup A_n is irreducible.

12. Let $T: G \to \mathrm{GL}(V)$ be a completely reducible finite-dimensional linear representation. Show that for every invariant subspace $U \subset V$ there is a decomposition $V = V_1 \oplus \dots \oplus V_m$ of V into a direct sum of minimal invariant subspaces and an $s \leq m$ such that $U = V_1 \oplus \dots \oplus V_s$.

13. Prove that the subspaces invariant under the representation Λ of $\mathrm{GL}_n(K)$ defined by formula (10) are precisely the left ideals in the ring of matrices of order n.

14.* Prove that the adjoint representation of the group $\mathrm{GL}(V)$ possesses only the two nontrivial invariant subspaces indicated in 1.6.

15.* Prove that $\mathrm{B}^+(V)$ and $\mathrm{B}^-(V)$ are minimal $\mathrm{GL}(V)$-invariant subspaces of the space $\mathrm{B}(V)$ of bilinear functions.

2. Complete Reducibility of Representations of Compact Groups

In this section we are concerned only with finite-dimensional representations.

2.1. One of the basic problems of representation theory is that of describing all representations of a given group (over a given field). In the preceding section we have seen that the description of completely reducible representations reduces to that of the irreducible representations (see also Section 3). Here we will show that all real or complex representations of finite groups are completely reducible. This result will subsequently be generalized to compact topological groups; this class includes, for example, the orthogonal group O_n.

The idea of the proof of complete reducibility is to equip the representation space with an inner product invariant under the action of the group. Then, given an arbitrary invariant subspace, one finds an invariant complement for it by taking its orthogonal complement.

2.2. We proceed to implement the program formulated above.

Definition. A real linear representation $T: G \to \mathrm{GL}(V)$ is called ORTHOGONAL if on the space V there is a positive definite symmetric bilinear function f INVARIANT under T.

The invariance of f means that

(1) $$f(T(g)x, T(g)y) = f(x, y)$$

for all $g \in G$ and all $x, y \in V$ or, equivalently, that

(2) $$T(g)_* f = f$$

for all $g \in G$, where the asterisk indicates the natural action of an invertible linear operator on bilinear functions (see Example 3, 1.6). Taking f as an inner product, we turn V into a *Euclidean space* in which the operators $T(g)$, for $g \in G$, are orthogonal.

Similarly, the complex linear representation $T: G \to \mathrm{GL}(V)$ is called UNITARY if on the space V there is a positive definite Hermitian sesquilinear function f invariant under T. Taking f as an inner product, we turn V into a *Hermitian space* in which the operators $T(g)$, for $g \in G$, are unitary.

Proposition. *Every orthogonal or unitary representation is completely reducible.*

PROOF. Let $T: G \to \mathrm{GL}(V)$ be an orthogonal representation of the group G. Let $U \subset V$ be an arbitrary invariant subspace. Denote by U^o the orthogonal complement of U with respect to an invariant inner product on V. It is known that

$$V = U \oplus U^o.$$

For each $g \in G$ the operator $T(g)$ is orthogonal and preserves U. By a well-known property, it preserves U^o as well. Hence, U^o is an invariant subspace complementing U.

The proof for a unitary representation is identical. □

2.3. Theorem 1. *Every real (complex) linear representation of a finite group is orthogonal (respectively, unitary).*

PROOF. Let $T: G \to \mathrm{GL}(V)$ be a real linear representation of a finite group G. Pick an arbitrary positive definite symmetric bilinear function f_0 on V, and construct a new symmetric bilinear function f by the rule

$$f = \sum_{h \in G} T(h)_* f_0. \tag{3}$$

Since

$$f(x, x) = \sum_{h \in G} f_0(T(h)^{-1}x, T(h)^{-1}x) > 0$$

for every nonnull vector $x \in V$, f is positive definite. We claim that f is invariant under T. In fact, for every $g \in G$,

$$T(g)_* f = \sum_{h \in G} T(g)_* T(h)_* f_0 = \sum_{h \in G} T(gh)_* f_0.$$

Since the equation $gx = h$ has a unique solution in G for every fixed h, the last sum above differs from the one in (3) only in the order of its terms. Hence $T(g)_* f = f$, as claimed.

The proof for a complex representation is identical. □

Corollary. *Every real or complex linear representation of a finite group is completely reducible.*

In point of fact, every linear representation of a finite group over a field whose characteristic does not divide the order of the group is completely reducible. (For a proof see, for example, [4].)

2.4. A TOPOLOGICAL GROUP is, by definition, a group endowed with a topology such that the group operations

$$x \mapsto x^{-1} \qquad \text{and} \qquad (x, y) \mapsto xy$$

are continuous maps.

Examples of topological groups.

1. Any group with the discrete topology.

2. $\mathrm{GL}(V)$, where V is an n-dimensional vector space over **R** or **C**. The topology is defined as on any (open) subset of the vector space $\mathrm{L}(V)$. That is, the continuous functions in this topology are exactly the continuous functions of the matrix elements (relative to some fixed basis, the choice of which, however, does not affect the topology). Since the matrix elements of the operators α^{-1} and $\alpha\beta$ are continuous functions of the matrix elements of α and β, $\mathrm{GL}(V)$ is indeed a topological group.

3. Any subgroup of a topological group endowed with the induced topology. In particular, every group of linear transformations of a real or complex vector space is a topological group.

A topological group is said to be COMPACT if it is compact as a topological space.

Examples of compact topological groups.

1. Any finite group endowed with the discrete topology.

2. The orthogonal group O_n.

3. The unitary group U_n.

4. Any closed subgroup of a compact topological group.

To prove the compactness of the groups O_n and U_n, we remind the reader of the following general fact: a subset of a real or complex vector space is compact if and only if it is closed and bounded. O_n is singled out in the space $\mathrm{L}_n(\mathbf{R})$ of real matrices by the algebraic equations

$$\sum_k a_{ik}a_{jk} = \delta_{ij}$$

and consequently is closed in $\mathrm{L}_n(\mathbf{R})$. The same equations yield the bounds $|a_{ij}| \leq 1$, which prove the boundedness of O_n in $\mathrm{L}_n(\mathbf{R})$. The compactness of the unitary group is established in an analogous manner.

A real or complex linear representation $T: G \to \mathrm{GL}(V)$ of the topological group G is said to be CONTINUOUS if it is a continuous map of the underlying topological spaces. This means that the matrix elements of the operator $T(g)$ depend continuously on g.

Examples.

1. Any real or complex linear representation of a discrete topological group.

2. If V is a real or complex vector space, then all representations of $\mathrm{GL}(V)$ considered in 1.6 are continuous.

For example, let us prove the continuity of the representation Φ (Example 3, 1.6). To this end we use its matrix expression (formula (14), 1.6). Let $a_{ij}, \tilde{a}_{ij}, x_{ij}$, and y_{ij} denote the elements of the matrices A, A^{-1}, X, and $\Phi(A)X$, respectively. Then

$$y_{ij} = \sum_{k,\ell} \tilde{a}_{ki}x_{k\ell}\tilde{a}_{\ell j}.$$

We see that $\Phi(A)$ is the linear transformation with coefficients (matrix elements) $\tilde{a}_{ki}\tilde{a}_{\ell j}$, which obviously depend continuously on the elements of the matrix A. This means precisely that Φ is a continuous representation.

One usually considers only *continuous* linear representations of topological groups. For this reason from now on we shall omit, as a rule, the adjective "continuous."

2.5. From the point of view of the theory of (continuous) linear representations, compact topological groups are similar to discrete ones. In particular, we have

Theorem 2. *Every real (complex) linear representation of a compact topological group is orthogonal (respectively, unitary).*

Recalling the proposition of 2.2, we derive the following

Corollary. *Every real or complex linear representation of a compact topological group is completely reducible.*

To prove Theorem 2 one can proceed as in the proof of Theorem 1, but now integration replaces summation over a finite group. It is known (see [7], for example) that on every compact topological group G one can define an INVARIANT INTEGRATION, meaning that to each continuous function f on G one can assign a number, denoted by $\int_G f(x)\,dx$, such that the mapping $f \mapsto \int_G f(x)\,dx$ possesses the following properties:

1) $\int_G (a_1 f_1(x) + a_2 f_2(x))\,dx = a_1 \int_G f_1(x)\,dx + a_2 \int_G f_2(x)\,dx$ (*linearity*);

2) if f is nonnegative everywhere and does not vanish identically, then we have $\int_G f(x)\,dx > 0$ (*positivity*);

3) $\int_G f(gx)\,dx = \int_G f(xg)\,dx = \int_G f(x)\,dx$ for every $g \in G$ (*invariance*).

Such an integration is *unique* up to a constant factor. Usually this factor is chosen so that

4) $\int_G 1\,dx = 1$.

In what follows we shall assume that this last condition is satisfied.

Examples.

1. The invariant integration on a finite group G is defined by the formula

$$\int_G f(x)\,dx = \frac{1}{|G|} \sum_{x \in G} f(x).$$

2. The invariant integration on the group $\mathbf{T} \simeq \mathrm{U}_1$ is defined by the formula

$$\int_G f(x)\,dx = \frac{1}{2\pi} \int_0^{2\pi} f(\mathrm{e}^{i\phi})\,d\phi.$$

3. In Chapter III we will show that the group SU_2 can be identified with the three-dimensional sphere in such a manner that the left and right translations are isometries of the sphere. Under this identification, the invariant integration on SU_2 can be defined as integration over the sphere with respect to the usual measure, multiplied by a factor of $(2\pi^2)^{-1}$.

4. Let N be a closed normal subgroup of the compact group G. The invariant integration on the quotient group G/N can be defined in terms of the invariant integration on G itself as

$$\int_{G/N} f(y)\,dy = \int_G f(xN)\,dx.$$

In this way one can, for example, define the invariant integration on the group SO_3, which, as we will see in Chapter III, is isomorphic to the quotient group $\mathrm{SU}_2/\{E, -E\}$.

The PROOF OF THEOREM 2 for a real linear representation $T\colon G \to \mathrm{GL}(V)$ of a compact group G proceeds as follows. Let f_0 be an arbitrary positive definite symmetric bilinear function on V. One defines a new symmetric bilinear function f by the rule

$$f(x, y) = \int_G f_0(T(g)x, T(g)y)\,dg \qquad (x, y \in V).$$

Next, using the properties of invariant integration one shows that f is positive definite and invariant. In the case of a complex representation one proceeds in the same manner, but one replaces symmetric bilinear functions by Hermitian sesquilinear functions. □

2.6. We now give *an alternate proof of Theorem 2* which does not resort to integration on the group.

We remark that on multiplying the sum in (3) by the factor $|G|^{-1}$ the function f becomes the *center of mass* of the finite set $M = \{T(h)_* f_0 \mid h \in G\}$ in the vector space $\mathrm{B}^+(V)$ of symmetric bilinear functions on V. For each $g \in G$ the transformation $T(g)_*$ maps the set M into itself (permuting its points in some way), and consequently preserves its center of mass.

Our proof of Theorem 2 will also rest on the idea of using the center of mass, but we must first replace the elementary definition, appropriate for the finite case, by the notion of center of mass of a compact set of positive measure.

Let V be a real vector space. Let $K \subset V$ be a compact set of positive measure. By definition, the CENTER OF MASS of K is the point (vector)

$$c(K) = \mu(K)^{-1} \int_K x\,\mu(dx). \tag{4}$$

Here x is a vector variable and μ denotes the usual measure on V; μ is defined to within a constant factor, but, as formula (4) shows, this freedom in the choice of μ does not affect the result $c(K)$.

The integral in (4) can be defined either coordinate-wise, or directly, as a limit of integral sums. The first definition proves its existence, while the second establishes its independence of the choice of a coordinate system (basis) in V.

We now show that

$$(5)\qquad c(\alpha K) = \alpha c(K) \qquad \text{for all } \alpha \in \mathrm{GL}(V).$$

In fact,

$$\begin{aligned} c(\alpha K) &= \mu(\alpha K)^{-1} \int_{\alpha K} x\mu(dx) \\ &= (\det \alpha \cdot \mu(K))^{-1} \int_K \alpha x \cdot \det \alpha \cdot \mu(dx) \\ &= \mu(K)^{-1} \alpha \int_K x\mu(dx) = \alpha c(K). \end{aligned}$$

(Moving α in front of the integral sign is permitted thanks to the continuity and linearity of the transformation α.)

Another important property is that *the center of mass of a compact set K lies in the convex hull of K.*

Recall that the CONVEX HULL of an arbitrary set $K \subset V$ is defined as

$$\operatorname{conv} K = \{\sum_{i=1}^{m} c_i x_i \mid x_i \in K,\ c_i \geq 0,\ \sum_{i=1}^{m} c_i = 1,\ m \text{ arbitrary}\}.$$

It is the smallest convex set containing K. One can show that *the convex hull of a compact set is closed* (see Appendix 3).

It follows from the definition of the integral that the center of mass $c(K)$ of the compact set K is a limit of vectors of the form

$$\mu(K)^{-1} \sum_{i=1}^{m} \mu(K_i) x_i,$$

where $x_i \in K_i \subset K$ and $\sum \mu(K_i) = \mu(K)$. Each such vector lies in $\operatorname{conv} K$, and since $\operatorname{conv} K$ is closed, $c(K) \in \operatorname{conv} K$, too.

2.7. PROOF OF THEOREM 2. Let $T\colon G \to \mathrm{GL}(V)$ be a real linear representation of the compact topological group G.

In the space $\mathrm{B}^+(V)$ of symmetric bilinear functions on V, consider the subset P of all positive definite functions. Obviously, P is closed under addition (the sum of two positive definite functions is again positive definite) and multiplication by positive numbers. This implies that P is convex. Moreover,

P is open, since in terms of matrices it is given by the condition that all principal minors be positive. Finally, it is plain that

$$(6) \qquad \alpha_* P \subset P$$

for all $\alpha \in \mathrm{GL}(V)$.

Let $K_0 \subset P$ be an arbitrary compact set of positive measure. Put

$$K = \bigcup_{h \in G} T(h)_* K_0$$

(cf. formula (3)). We claim that the set K enjoys the following properties:

(K1) $K \subset P$;

(K2) $T(g)_* K = K$ for all $g \in G$;

(K3) K is compact.

Property (K1) is a consequence of (6). (K2) follows from the equality

$$T(g)_* T(h)_* K_0 = T(gh)_* K_0.$$

To prove (K3), consider an arbitrary sequence $T(h_n)_* f_n$ ($h_n \in G, f_n \in K_0$) of elements of K. Since G and K_0 are compact, we can, passing to a subsequence if necessary, ensure that $h_n \to h \in G$ and $f_n \to f \in K_0$. Then $T(h_n)_* f_n \to T(h)_* f \in K$ (here we used the continuity of the representation T).

Now consider the center of mass $f = c(K)$ of K in the space $\mathrm{B}^+(V)$. Since $c(K) \in \operatorname{conv} K$ (see 2.6), (K1) and the convexity of P guarantee that $f \in P$, i.e., f is a positive definite symmetric bilinear function. Properties (K2) and (5) imply that $T(g)_* f = f$ for all $g \in G$, i.e., f is G-invariant. Thus, T is an orthogonal representation, as asserted.

The complex version of the theorem is proved in an analogous manner, with the difference that instead of $\mathrm{B}^+(V)$ one works with the space $\mathrm{H}^+(V)$ of positive definite Hermitian sesquilinear functions. Notice that $\mathrm{H}^+(V)$ is a real (and not complex) vector space, and the notion of center of mass introduced in 2.6 can be used in the indicated proof with no modifications.

Questions and Exercises

1. Let T be an orthogonal or unitary representation of the group G. Prove that all complex eigenvalues of the operators $T(g)$, $g \in G$, have modulus one.

2. Give an example of a nonunitary complex representation of $\mathbf{Z}$.

3. Let T be a real representation of $\mathbf{Z}_3$ in which the generator of $\mathbf{Z}_3$ goes into the linear operator with the matrix $\begin{pmatrix} 0 & -1 \\ 1 & -1 \end{pmatrix}$. Find a positive definite symmetric bilinear function invariant under T.

4. Which of the following topological groups are compact: $\mathbf{Z}$, $\mathbf{Z}_m$, $\mathbf{T}$, $\mathrm{SL}_n(\mathbf{R})$?

5. Let $T: G \to \mathrm{GL}(V)$ be a continuous real or complex representation of the topological group G, and let $U \subset V$ be an invariant subspace. Show that the representations T_U and $T_{V/U}$ are continuous.

6. Let V be a real or complex vector space. Show that the adjoint representation of the group GL(V) (Example 2, 1.6) is continuous.

7.* Let K_1 and K_2 be compact sets of positive measure in the real vector space V. Prove that

a) if the symmetric difference of K_1 and K_2 has measure zero, then $c(K_1) = c(K_2)$;

b) if $\mu(K_1 \cap K_2) = 0$, then $c(K_1 \cup K_2)$ lies on the segment connecting $c(K_1)$ and $c(K_2)$.

8.* Give an example of a compact subset of positive measure in the set P of positive definite symmetric bilinear functions.

3. Basic Operations on Representations

Various methods of obtaining new representations from one or more other representations play an important role in representation theory. We have already encountered certain constructions of this sort, namely, composing a representation and a homomorphism (0.10), and passing to a subrepresentation or quotient representation (1.2). In this section we consider a number of other constructions that will be needed later.

3.1. The Contragredient or Dual Representation. Given any linear representation $T: G \to \mathrm{GL}(V)$, we define in a canonical manner the CONTRAGREDIENT or DUAL representation $T': G \to \mathrm{GL}(V')$ in the dual space V' of V. (Recall that the elements of V' are the linear functions on V.)

Definition. $(T'(g)f)(x) = f(T(g)^{-1}x) \quad (g \in G,\ f \in V',\ x \in V)$.

T' is a subrepresentation of the representation T_* of G in the space of all K-valued functions on V (see 0.9).

For a finite-dimensional representation T, the contragredient representation can be described in terms of matrices as follows. Let $(e) = (e_1, \dots, e_n)$ and $(\varepsilon) = (\varepsilon_1, \dots, \varepsilon_n)$ be a basis in V and the dual basis in V' respectively, i.e., $\varepsilon_i(e_j) = \delta_{ij}$. Let

$$T_{(e)}(g) = [a_{ij}] \qquad \text{and} \qquad T'_{(\varepsilon)}(g) = [b_{ij}].$$

According to the definition,

$$(T'(g)\varepsilon_i)(T(g)e_j) = \varepsilon_i(e_j) = \delta_{ij}.$$

Since

$$T'(g)\varepsilon_i = \sum_k b_{ki}\varepsilon_k, \qquad T(g)e_j = \sum_\ell a_{\ell j}e_\ell,$$

we have that

$$(T'(g)\varepsilon_i)(T(g)e_j) = \sum_k b_{ki}a_{kj} = \delta_{ij}.$$

This means that $(T'_{(\varepsilon)}(g))'T_{(e)}(g) = E$ or, equivalently,

$$T'_{(\varepsilon)}(g) = ((T_{(e)}(g))')^{-1}. \tag{1}$$

In particular, if T is an orthogonal representation and the basis (e) is orthonormal (with respect to an invariant inner product), then $T_{(e)}(g)$ is an orthogonal matrix, and so $((T_{(e)}(g))')^{-1} = T_{(e)}(g)$. In this case $T'_{(\varepsilon)}(g) = T_{(e)}(g)$, and so $T' \simeq T$.

If T is a unitary representation and the basis (e) is orthonormal, then

$$T'_{(\varepsilon)}(g) = \overline{T_{(e)}(g)}, \tag{2}$$

where the bar denotes complex conjugation.

It also follows from formula (1) that $T'' \simeq T$ for every representation T.

Theorem 1. *Let T be an irreducible finite-dimensional representation. Then T' is irreducible.*

Proof. Let $U \subset V'$ be a T'-invariant subspace. Consider its annihilator

$$U^0 = \{x \in V \mid f(x) = 0 \quad \text{for all } f \in U\} \subset V.$$

It is a T-invariant subspace. In fact, for any $g \in G$, $x \in U^0$, and $f \in U$ we have

$$f(T(g)x) = (T'(g)^{-1}f)(x) = 0,$$

because $T'(g)^{-1}f \in U$. It is known from the theory of systems of linear equations that $\dim U^0 = \dim V - \dim U$. Since T is irreducible, U^0 is equal to 0 or V, and correspondingly U is equal to V' or 0. This means that V' contains no nontrivial T'-invariant subspaces, as we needed to show. □

3.2 Sums of Representations. Let

$$T_1: G \to \mathrm{GL}(V_1) \qquad \text{and} \qquad T_2: G \to \mathrm{GL}(V_2)$$

be two linear representations of the group G.

Definition. The SUM OF T_1 AND T_2 is the representation $T_1 + T_2$ of G in the space $V_1 \oplus V_2$ defined by the rule

$$(T_1 + T_2)(g)(x_1 + x_2) = T_1(g)x_1 + T_2(g)x_2$$
$$(g \in G,\ x_1 \in V_1,\ x_2 \in V_2).$$

The sum of an arbitrary finite number of representations is defined in a similar manner. A sum of representations is independent, up to an isomorphism, of the order of its summands.

This definition makes it clear that the spaces V_1 and V_2, canonically imbedded in $V_1 \oplus V_2$, are invariant under $T_1 + T_2$. Conversely, if the space V of a representation T of G can be written as the direct sum of two T-invariant subspaces V_1 and V_2, then T coincides with the sum of the representations T_{V_1} and T_{V_2}. In fact,

$$T(g)(x_1 + x_2) = T(g)x_1 + T(g)x_2 = T_{V_1}(g)x_1 + T_{V_2}(g)x_2$$

for all $x_1 \in V_1$, $x_2 \in V_2$, which is precisely the definition of the sum of the representations T_{V_1} and T_{V_2}. Analogous assertions are of course true for a sum of finitely many representations.

In terms of matrices, the sum T of the representations $T_i: G \to \mathrm{GL}(V_i)$, for $i = 1, 2, \ldots, m$, is described as follows. Let (e) be a basis in $V = V_1 \oplus V_2 \oplus \ldots \oplus V_m$ that is a union of bases $(e)_i$ in V_i. Then in block form

$$T(g)_{(e)} = \begin{pmatrix} T(g)_{(e)_1} & & & 0 \\ & T(g)_{(e)_2} & & \\ & & \ddots & \\ 0 & & & T(g)_{(e)_m} \end{pmatrix}.$$

The notion of a sum of representations is suitable for formulating properties of completely reducible representations.

Theorem 2. *Every completely reducible finite-dimensional linear representation is isomorphic to a sum of irreducible representations. Conversely, every sum of irreducible representations is completely reducible.*

This is simply a reformulation of Theorem 2 and of the first assertion of Theorem 3 of 1.5. □

Theorem 3. *Suppose the representation $T: G \to \mathrm{GL}(V)$ is isomorphic to a sum of irreducible representations $T_i: G \to \mathrm{GL}(V_i)$, $i = 1, \dots, m$. Then every subrepresentation of T as well as every quotient representation of T is isomorphic to a sum of some of the representations T_i.*

PROOF. It suffices to prove the assertion for quotient representations, since every subrepresentation T_U of T is isomorphic to the quotient representation $T_{V/W}$, where W is an invariant complement of the subspace U.

Let U be an invariant subspace. By Theorem 3 of 1.5, it admits a complement of the form $V_{i_1} \oplus \dots \oplus V_{i_p}$, and then $T_{V/U} \simeq T_{V_{i_1} \oplus \dots \oplus V_{i_p}} \simeq T_{i_1} \oplus \dots \oplus T_{i_p}$. □

Corollary. *Let $T: G \to \mathrm{GL}(V)$ be a linear representation. Let $V_1, \dots, V_m$ be minimal invariant subspaces such that the representations $T_i = T_{V_i}$ are pairwise nonisomorphic. Then $V_1, \dots, V_m$ are linearly independent.*

PROOF. Suppose this is not the case. Then there is a $k < m$ such that the subspaces $V_1, \dots, V_k$ are linearly independent, whereas $V_{k+1} \cap \sum_{i=1}^{k} V_i = \Delta \neq 0$. Since $\Delta \subset V_{k+1}$ and V_{k+1} is a minimal invariant subspace, $\Delta = V_{k+1}$, i.e., $V_{k+1} \subset \sum_{i=1}^{k} V_i$. But then, by Theorem 3, T_{k+1} is isomorphic to one of the representations $T_1, \dots, T_k$, which contradicts the hypothesis. □

We show next that the decomposition of a completely reducible representation into a sum of irreducible components is, in a certain sense, unique.

Theorem 4. *Let T be a linear representation. If*

$$T \simeq T_1 + \dots + T_m \simeq S_1 + \dots + S_p,$$

where T_i and S_j are irreducible representations, then $m = p$ and, for a suitable labeling, $T_i \simeq S_i$.

(Compare this result with the theorem asserting the uniqueness of the decomposition of a positive integer into prime factors.)

PROOF. By hypothesis, the representation space V of T admits two decompositions into a direct sum of minimal invariant subspaces,

$$V = V_1 \oplus \dots \oplus V_m = U_1 \oplus \dots \oplus U_p,$$

such that $T_{V_i} \simeq T_i$ and $T_{U_j} \simeq S_j$.

The proof proceeds by induction on m. Applying Theorem 3 of 1.5 to the invariant subspace $U = U_1$, we deduce that $V = U \oplus V_{i_1} \oplus \dots \oplus V_{i_k}$ for certain $i_1, \dots, i_k$. Then

$$S_1 \simeq T_U \simeq T_{V/(V_{i_1} \oplus \dots \oplus V_{i_k})} \simeq T_{j_1} + \dots + T_{j_\ell},$$

where $\{j_1, \dots, j_\ell\} = \{1, \dots, m\} \setminus \{i_1, \dots, i_k\}$. Since the representation S_1 is irreducible, $\ell = 1$. Now let us relabel the representations T_i so that $j_1 = 1$. Then $S_1 \simeq T_1$ and

$$V = U \oplus V_2 \oplus \dots \oplus V_m.$$

This says that

$$T_{V/U} \simeq T_2 + \dots + T_m.$$

On the other hand, it is clear that

$$T_{V/U} \simeq S_2 + \dots + S_p.$$

Applying the inductive hypothesis to $T_{V/U}$, we conclude that $m = p$ and, after a suitable relabeling, $T_i \simeq S_i$ for all $i \geq 2$. Since $T_1 \simeq S_1$, the assertion of the theorem is also true for T. □

We remark that Theorem 4 does not imply the uniqueness of the decomposition of the representation space into a direct sum of minimal invariant subspaces. Such uniqueness does not hold, as can be seen even in the case of a trivial representation (see the end of 1.5).

3.3. Products of Representations. Let $T: G \to \mathrm{GL}(V)$ and $S: G \to \mathrm{GL}(U)$ be two linear representations of the group G.

Definition. The PRODUCT OF T AND S is the representation TS of G in the space $V \otimes U$ defined by the rule

$$TS(g) = T(g) \otimes S(g).$$

(For the definitions of the tensor product for vector spaces and linear operators, see Appendix 2.)

Sometimes TS is referred to as the tensor product of the representations T and S. However, we reserve this term for another notion, defined below in 3.4.

Let us give the matrix interpretation of the product of finite-dimensional representations. To this end we pick bases

$$(e) = (e_1, \dots, e_n) \subset V \qquad \text{and} \qquad (f) = (f_1, \dots, f_m) \subset U$$

of the spaces V and U respectively. Each element $x \in V \otimes U$ can be uniquely expressed as

$$x = \sum x_{ij}(e_i \otimes f_j).$$

How is the matrix $X = [x_{ij}]$ transformed under the action of the operator $TS(g)$ on x?

Let $T_{(e)}(g) = [a_{ij}]$ and $S_{(f)}(g) = [b_{ij}]$. Then

$$T(g)e_i = \sum_k a_{ki}e_k, \qquad S(g)f_j = \sum_\ell b_{\ell j}f_\ell,$$

and

$$\begin{aligned} TS(g)x &= \sum_{i,j} x_{ij}(T(g)e_i \otimes S(g)f_j) \\ &= \sum_{i,j,k,\ell} x_{ij}a_{ki}b_{\ell j}(e_k \otimes f_\ell) \\ &= \sum_{i,j}(\sum_{k,\ell} a_{ik}x_{k\ell}b_{j\ell})(e_i \otimes f_j). \end{aligned}$$

Hence, X transforms according to the rule

$$X \mapsto T_{(e)}(g)XS_{(f)}(g)'.$$

We thus obtain the following matrix interpretation of the product of two representations:

(3) $$TS(g)X = T(g)XS(g)' \qquad (X \in \mathrm{L}_{n,m}(K)).$$

(Here T and S are regarded as matrix representations, and the representation space of TS is interpreted as the space of $(n \times m)$-matrices.)

Examples.

1. Let T be an n-dimensional linear representation of a group G in the space V, and $I = I_m$ the m-dimensional trivial representation of G in the space U. Let us examine the representation TI.

In terms of matrices, TI is given by the formula

$$TI(g)X = T(g)X \qquad (X \in \mathrm{L}_{n,m}(K)).$$

When $U = V'$, this coincides with the composition of the representation Λ of $\mathrm{GL}(V)$ considered in Example 1 of 1.6 and the representation $T: G \to \mathrm{GL}(V)$. In the general case one can show, proceeding exactly as in 1.6, that

$$TI_m \simeq mT \quad (= \underbrace{T + \ldots + T}_{m \text{ times}}).$$

The corresponding decomposition of $V \otimes U$ into a direct sum of invariant subspaces is readily described in invariant terms as well. It has the form

$$V \otimes U = (V \otimes f_1) \oplus \ldots \oplus (V \otimes f_m),$$

where $(f_1, \ldots, f_m)$ is a basis of U. The map

$$x \mapsto x \otimes f_i$$

is an isomorphism of the representations T and $(TI)_{V\otimes f_i}$.

2. The representation TT' is, according to (1) and (3), described in terms of matrices as

$$TT'(g)X = T(g)XT(g)^{-1} \qquad (X \in \mathrm{L}_n(K)). \tag{5}$$

This shows that $TT' = \mathrm{Ad}\circ T$, where Ad is the adjoint representation of $\mathrm{GL}(V)$ (Example 2, 1.6).

3. The representation $(T')^2 = T'T'$ is given in terms of matrices by the formula

$$(T')^2(g)X = T(g)'^{-1}XT(g)^{-1} \qquad (X \in \mathrm{L}_n(K)). \tag{6}$$

Consequently, $(T')^2 = \Phi\circ T$, where Φ is the natural representation of $\mathrm{GL}(V)$ in the space $\mathrm{B}(V) = V' \otimes V'$ (see Example 3, 1.6).

4. In the case where one of the representations T, S is *one-dimensional*, the product TS has a particularly simple meaning. Suppose, for example, that $T: G \to \mathrm{GL}(V)$ is an arbitrary representation of the group G, and $S: G \to \mathrm{GL}(U)$ is a one-dimensional representation, i.e., a homomorphism of G into K^*. Pick a nonnull vector $u_0 \in U$. The map

$$\sigma: x \mapsto x \otimes u_0$$

is an isomorphism of V onto $V \otimes U$. For every $g \in G$ we have

$$TS(g)\sigma x = T(g)x \otimes S(g)u_0 = S(g)T(g)x \otimes u_0 = \sigma S(g)T(g)x,$$

where $S(g)T(g)$ is understood as the product of the operator $T(g)$ by the scalar $S(g)$. Thus TS is isomorphic to the representation

$$g \mapsto S(g)T(g) \in GL(V).$$

The product of an arbitrary finite number of representations is defined in a natural manner. In particular, if T is a representation of G in a vector space V, then $T^kT'^\ell$ is a representation of G in the space of tensors of type (k, ℓ) over V. Such representations are often encountered in mathematical and physical applications of representation theory.

Examples 2 and 3 (see also the corresponding examples in 1.6) show that a product of irreducible representations is not necessarily irreducible. Decomposing such a product into a sum of irreducible representations is one of the most important problems of representation theory.

3.4. Tensor Products of Representations of Two Groups. Let $T: G \to \mathrm{GL}(V)$ and $S: H \to \mathrm{GL}(U)$ be two representations of the groups G and H.

Definition. The TENSOR PRODUCT OF T AND S is the representation $T \otimes S$ of the group $G \times H$ in the space $V \otimes U$, defined by the rule

$$(T \otimes S)(g, h) = T(g) \otimes S(h)$$

(here we should really write $(T \otimes S)((g, h))$!).

In the matrix interpretation (cf. 3.3), the tensor product of two finite-dimensional representations takes the form

$$(T \otimes S)(g, h)X = T(g)XS(h)'. \tag{7}$$

Here, in contrast to formula (3), the factors on the left and right of the matrix X are independent (even if $G = H$).

Let i_1 denote the canonical imbedding of the group G in $G \times H$, i.e., $i_1(g) = (g, e)$. Then $(T \otimes S)\circ i_1$ is a representation of G. It is clear from the definition that $(T \otimes S)\circ i_1 = TI$, where I is the trivial representation of G in U. Therefore (see Example 1 of 3.3),

$$(T \otimes S)\circ i_1 \simeq (\dim U)T.$$

Similarly, if i_2 denotes the canonical imbedding of H in $G \times H$, then

$$(T \otimes S)\circ i_2 = IS,$$

where I is now the trivial representation of H in V. Therefore,

$$(T \otimes S)\circ i_2 \simeq (\dim V)S.$$

A very important example. Let T be a representation of the group G in the n-dimensional space V. Consider the representation $T \otimes T'$ of $G \times G$ in $V \otimes V'$. In terms of matrices it is described as

$$(T \otimes T')(g_1, g_2)X = T(g_1)XT(g_2)^{-1} \qquad (X \in \mathrm{L}_n(K)) \tag{8}$$

(cf. Example 2, 3.3).

If one uses the canonical identification of the spaces $V \otimes V'$ and $\mathrm{L}(V)$ (see Appendix 2), then $T \otimes T'$ can be described in invariant form as

$$(T \otimes T')(g_1, g_2)\xi = T(g_1)\xi T(g_2)^{-1} \qquad (\xi \in \mathrm{L}(V)). \tag{9}$$

This follows from (8). In fact, the matrix assigned to a linear operator when that operator is viewed as an element of the tensor product $V \otimes V'$ coincides with its usual matrix (see Appendix 2), and to the product of matrices there corresponds the product of linear operators.

3.5. Extension of the Ground Field. Let K' be an extension of the field K.

The group $\mathrm{GL}_n(K)$ is then a subgroup of $\mathrm{GL}_n(K')$. Consequently, every n-dimensional matrix representation T of an arbitrary group G over K can also be regarded as an n-dimensional representation of G *over* K', and in this capacity we denote it by $E^K_{K'}T$. In exact terms, $E^K_{K'}T$ is the composition of the canonical imbedding of $\mathrm{GL}_n(K)$ in $\mathrm{GL}_n(K')$ with the representation T.

A similar operation can be defined for linear representations.

First of all, every vector space V over K can be included in a vector space $E^K_{K'}V$ over K' in such a way that a basis (e) of V is simultaneously a basis (over K') of $E^K_{K'}V$. Accordingly, every linear transformation α of V extends to a linear transformation $E^K_{K'}\alpha$ of $E^K_{K'}V$ that, in the basis (e), has the same matrix as α:

$$(E^K_{K'}\alpha)_{(e)} = \alpha_{(e)}.$$

This yields an imbedding $\mathrm{L}(V) \subset \mathrm{L}(E^K_{K'}V)$, which in turn induces a group imbedding

$$\mathrm{GL}(V) \subset \mathrm{GL}(E^K_{K'}V).$$

Now let T be a linear representation of G in V. Setting

$$(E^K_{K'}T)(g) = E^K_{K'}T(g),$$

we obtain a linear representation of G in $E^K_{K'}V$.

3.6. Let us examine in more detail the most important case for applications: $K = \mathbf{R}$, $K' = \mathbf{C}$. The operation $E^{\mathbf{R}}_{\mathbf{C}}$ is called COMPLEXIFICATION (of vector spaces, linear operators, and representations). For simplicity, we shall denote it by the index $\mathbf{C}$, writing $T_{\mathbf{C}}$, for example, instead of $E^{\mathbf{R}}_{\mathbf{C}}T$.

One reason why complexification is often useful is that in a complex vector space, in contrast to a real one, every linear operator has an eigenvector.

Example. By complexifying the representation of $\mathbf{R}$ through rotations of the Euclidean plane (Example 1, 0.7), we are allowed to write it in the form

$$t \mapsto \begin{pmatrix} \mathrm{e}^{it} & 0 \\ 0 & \mathrm{e}^{-it} \end{pmatrix}.$$

To do this we must pass from an orthonormal basis (e_1, e_2) of the Euclidean plane to the new basis $(e_1 - ie_2, e_1 + ie_2)$.

Theorem 5. *Two finite-dimensional real linear representations are isomorphic if and only if their complexifications are isomorphic.*

PROOF. Let T_1 and T_2 be n-dimensional real representations of the group G. In matrix form, the fact that T_1 and T_2 are isomorphic means that there exists a real matrix C satisfying the following two conditions:

(C1) $CT_1(g) = T_2(g)C$ for all $g \in G$;

(C2) $\det C \neq 0$.

Similarly, the fact that the representations $(T_1)_{\mathbf{C}}$ and $(T_2)_{\mathbf{C}}$ are isomorphic means that there exists a complex matrix C satisfying the same two conditions. This clearly proves the implication $(T_1 \simeq T_2) \Rightarrow ((T_1)_{\mathbf{C}} \simeq (T_2)_{\mathbf{C}})$.

To prove the converse implication, we remark that condition (C1) is in fact a homogeneous system of linear equations with real coefficients for the entries of the matrix C. Its general solution has the form $t_1C_1 + t_2C_2 + \ldots + t_mC_m$, where $C_1, \ldots, C_m$ are linearly independent real matrices. The determinant $\det(t_1C_1 + t_2C_2 + \ldots + t_mC_m)$ is a polynomial d in $t_1, \ldots, t_m$ with real coefficients.

Suppose now that $(T_1)_{\mathbf{C}} \simeq (T_2)_{\mathbf{C}}$. Then there exist complex numbers $\tau_1, \ldots, \tau_m$ such that $d(\tau_1, \ldots, \tau_m) \neq 0$, and so d is not the zero polynomial. But in this case there also exist real numbers $\tau'_1, \ldots, \tau'_m$ such that $d(\tau'_1, \ldots, \tau'_m) \neq 0$. Therefore, $T_1 \simeq T_2$, as needed. □

3.7. What can be said about the connection between the invariant subspaces of the representation $T: G \to \mathrm{GL}(V)$ and those of its complexification $T_{\mathbf{C}}: G \to \mathrm{GL}(V_{\mathbf{C}})$? Obviously, *the complexification $U_{\mathbf{C}}$ of any T-invariant subspace $U \subset V$ is a $T_{\mathbf{C}}$-invariant subspace.* However, $V_{\mathbf{C}}$ may contain invariant subspaces which do not arise in this manner. For instance, in the example of 3.6, the representation T is irreducible, whereas $T_{\mathbf{C}}$ possesses one-dimensional invariant subspaces.

To answer the question posed above, we introduce the operation of *complex conjugation* in the space $V_{\mathbf{C}}$. Each vector $z \in V_{\mathbf{C}}$ can be uniquely written as $z = x + iy$, with $x, y \in V$. Put $\bar{z} = x - iy$. In a basis consisting of real vectors (i.e., vectors in V), the coordinates of $\bar{z}$ are the complex conjugates of the coordinates of z.

Complex conjugation is an anti-linear transformation, that is, $\overline{z + u} = \bar{z} + \bar{u}$ and $\overline{cz} = \bar{c}\bar{z}$ for $c \in \mathbf{C}$. It follows that it transforms every subspace of $V_{\mathbf{C}}$ into a subspace of the same dimension.

Lemma. *The subspace $W \subset V_{\mathbf{C}}$ is the complexification of some subspace $U \subset V$ if and only if $\overline{W} = W$.*

Proof. It is plain that if $W = U_{\mathbf{C}}$, then $\overline{W} = W$. Conversely, suppose that $\overline{W} = W$. Then the subspace W contains, together with each vector $z = x + iy \quad (x, y \in V)$, the vector $\bar{z} = x - iy$, and hence also the linear combinations $x = \frac{1}{2}(z + \bar{z})$ and $y = \frac{1}{2}(z - \bar{z})$. Consequently, $W = U_{\mathbf{C}}$, where $U = W \cap V$. □

Since the operators $T_{\mathbf{C}}(g)$, for $g \in G$, take real vectors into real vectors, they commute with complex conjugation:

$$T_{\mathbf{C}}(g)\bar{z} = \overline{T_{\mathbf{C}}(g)z} \qquad (z \in V_{\mathbf{C}}). \tag{10}$$

Therefore, $\overline{W}$ is an invariant subspace whenever W is invariant. Consider the subspaces $W + \overline{W}$ and $W \cap \overline{W}$. They are also G-invariant. Moreover, they coincide with their complex conjugates:

$$\overline{W + \overline{W}} = \overline{W} + W = W + \overline{W}, \qquad \overline{W \cap \overline{W}} = \overline{W} \cap W = W \cap \overline{W}.$$

By the preceding lemma, $W + \overline{W}$ and $W \cap \overline{W}$ are complexifications of G-invariant subspaces of V.

Theorem 6. *Let $T: G \to \mathrm{GL}(V)$ be an irreducible real linear representation. Then $T_{\mathbf{C}}$ is either irreducible or the sum of two irreducible representations. In the second case $V_{\mathbf{C}}$ decomposes into the direct sum of two complex-conjugate minimal invariant subspaces.*

Proof. Let W be a minimal invariant subspace of $V_{\mathbf{C}}$. Then $W + \overline{W}$ is the complexification of an invariant subspace of V, which must coincide with V in view of the irreducibility of the representation T. Hence $W + \overline{W} = V_{\mathbf{C}}$.

Now consider the invariant subspace $W \cap \overline{W} \subset W$. It must either coincide with W or be the null subspace. In the first case, $W = \overline{W} = V_{\mathbf{C}}$ and the representation $T_{\mathbf{C}}$ is irreducible. In the second case, $V_{\mathbf{C}} = W \oplus \overline{W}$, and $T_{\mathbf{C}}$ decomposes into the sum of two irreducible representations. □

Examples.

1. In the example considered in 3.6, the second alternative of Theorem 6 holds true.

2. In the Euclidean plane, consider an equilateral triangle $A_1A_2A_3$ centered at the origin. For each permutation $\sigma \in S_3$ we let $T(\sigma)$ denote the orthogonal transformation that takes the vertex A_i into $A_{\sigma(i)}$ $(i = 1, 2, 3)$. $T(\sigma)$ is either

the identity transformation, or the rotation by $2\pi/3$ in one of the two possible directions, or the reflection in one of the altitudes of the triangle $A_1A_2A_3$. We thus get a faithful two-dimensional real representation T of the group S_3. It is obviously irreducible.

Using Theorem 6 it is readily established that the complexification $T_{\mathbf{C}}$ is also irreducible. Otherwise it would decompose into the sum of two one-dimensional representations, and in a suitable basis all the operators $T_{\mathbf{C}}(\sigma)$, $\sigma \in S_3$, would be given by diagonal matrices. The latter is impossible, however, since diagonal matrices commute with one another, whereas the group S_3 is not commutative.

Assertions similar to Theorems 5 and 6 permit us to reduce all questions concerning real representations to questions concerning complex representations. Since complex representations are simpler to describe than real ones, they constitute the main object of representation theory.

3.8. Lifting and Factoring. In 0.10 we considered the composition of a linear transformation and a homomorphism. A particular case of that construction is the composition $S \circ p$ of a linear representation

$$S: G/N \to \mathrm{GL}(V)$$

of a quotient group G/N and the canonical homomorphism

$$p: G \to G/N.$$

We call it the LIFT OF THE REPRESENTATION S. The representation $S \circ p$ has the property that its kernel contains the subgroup N:

$$(S \circ p)(h) = \varepsilon \qquad \text{for all } h \in N.$$

Conversely, every linear representation T of G whose kernel contains N takes all elements of a given coset of N in G into the same operator and so can be "factored" through p, i.e., $T = S \circ p$, where S is a linear representation of the quotient group G/N. We call the transition from T to S FACTORING THE REPRESENTATION T WITH RESPECT TO THE SUBGROUP N.

We thus establish a *one-to-one correspondence between the linear representations of the quotient group G/N and those linear representations of G whose kernel contains N.*

Examples.

1. $G = \mathrm{GL}_n(K)$, $N = \mathrm{SL}_n(K)$. Since N is the kernel of the epimorphism

$$\det: \mathrm{GL}_n(K) \to K^*,$$

$G/N \simeq K^*$. Every linear representation S of K^* can be lifted to yield a linear representation T of $\mathrm{GL}_n(K)$. For example, if $S: t \mapsto t^m$ (a one-dimensional representation), then $T: A \mapsto (\det A)^m$.

2. $G = S_4$, $N = \{\varepsilon, (12)(34), (13)(24), (14)(23)\}$. ($N$ is called Klein's four-group.) Each coset of N in G contains exactly one permutation that keeps 1 fixed; hence, $G/N \simeq S_3$. This observation can be used to build a linear representation of S_4 from any given linear representation of S_3.

In particular, from the two-dimensional irreducible representation of S_3 constructed in Example 2 of 3.7 one obtains a two-dimensional irreducible representation of S_4.

3. All linear representations of the group $\mathbf{Z}_m = \mathbf{Z}/m\mathbf{Z}$ are obtained by factoring linear representations T of $\mathbf{Z}$ with the property

$$T(m) = T(1)^m = \varepsilon.$$

3.9. The considerations of 3.8 apply to the description of *one-dimensional linear representations.*

Let T be a one-dimensional representation of G. Then

$$T(ghg^{-1}h^{-1}) = T(g)T(h)T(g)^{-1}T(h)^{-1} = 1$$

for all $g, h \in G$. Consequently, $\operatorname{Ker} T$ contains the subgroup of G generated by all commutators $(g, h) = ghg^{-1}h^{-1}$. The latter is called the COMMUTATOR SUBGROUP OF G and is denoted (G, G). It is a normal subgroup of G, since the set of all commutators is invariant under any (in particular, any inner) automorphism a of G:

$$a((g, h)) = (a(g), a(h)).$$

(Recall that a normal subgroup is by definition a subgroup invariant under all inner automorphisms.)

Therefore, *every one-dimensional representation of the group G is the lift of a one-dimensional representation of the quotient group $G/(G,G)=A(G)$.* We remark that $A(G)$ is *abelian*. In fact, let p denote the canonical homomorphism of G onto $A(G)$. Then for any $g, h \in G$,

$$(p(g), p(h)) = p((g, h)) = 1,$$

whence $p(g)p(h) = p(h)p(g)$.

Example. Let us find all one-dimensional representations of the symmetric group S_n. To this end we compute its commutator subgroup. Since the commutator of any two permutations is an even permutation, $(S_n, S_n) \subset A_n$. We show that $(S_n, S_n) = A_n$.

It is a straightforward matter to check that the commutator of the transpositions (ik) and (jk) (with distinct i, j, k) is the triple cycle (ijk). Let $H \subset S_n$ denote the subgroup generated by all triple cycles. Using a permutation of the form (ijk) one can take 1 to any prescribed element of the set $\{1, 2, \dots, n\}$; then, using a permutation of the form $(2jk)$, one can take 2 to any prescribed element of $\{1, 2, \dots, n\}$ while keeping 1 in its place, and so on, up to and including $n-2$. This shows that for every $\sigma \in A_n$ there exists an $\eta \in H$ such that $\eta(i) = \sigma(i)$ for $i = 1, 2, \dots, n-2$. Since η and σ have the same parity, one also has that $\eta(n-1) = \sigma(n-1)$ and $\eta(n) = \sigma(n)$. Hence, $H = A_n$, and since $(S_n, S_n) \supset H$, we conclude that $(S_n, S_n) = A_n$.

The quotient group $S_n/A_n \simeq \mathbf{Z}_2$ has two one-dimensional representations: one trivial, and the other taking the generator to -1. To the first there corresponds the trivial one-dimensional representation I of S_n, while to the second there corresponds the representation Π which takes all even permutations to 1 and all odd permutations to -1.

Questions and Exercises

1. Describe the dual of a trivial representation.

2. Prove that if the representation T' is irreducible, then so is T.

3. Prove that $(R+S)' \simeq R' + S'$ for any two representations R and S of a group G.

4. Prove that if the representation T is completely reducible, then so is T'.

5. Prove that the identity representation of $\mathrm{SL}_2(K)$ is isomorphic to its dual.

6. Let $T: G \to \mathrm{GL}(V)$ be a completely reducible representation, and let $U \subset V$ be an invariant subspace. Show that $T \simeq T_U + T_{V/U}$.

7. Let T_1, T_2, and S be completely reducible finite-dimensional linear representations. Show that if $T_1 + S \simeq T_2 + S$, then $T_1 \simeq T_2$.

8. Prove that $(T_1 + T_2)S \simeq T_1S + T_2S$ for any representations T_1, T_2, and S of G.

9. Prove that $TS \simeq ST$ for any representations T and S of G.

10. Describe the square of a representation in terms of matrices.

11.* Let V and U be complex vector spaces, and let $\alpha \in \mathrm{L}(V)$, $\beta \in \mathrm{L}(U)$. The product of the representations $t \mapsto e^{t\alpha}$ and $t \mapsto e^{t\beta}$ of $\mathbf{C}$ is necessarily of the form $t \mapsto e^{t\gamma}$, where $\gamma \in \mathrm{L}(V \otimes U)$. Find the operator γ.

12. Let T and S be an irreducible representation and a one-dimensional representation, respectively, of the group G. Show that TS is irreducible.

13. Prove formula (9) without resorting to the matrix interpretation.

14. Interpret the representation $T \otimes T$ in terms of matrices, and compare it with T^2.

15. Prove that the complexification of any odd-dimensional irreducible real representation is irreducible.

16. Find all finite-dimensional representations of O_n whose kernels contain SO_n.

17. Find all one-dimensional representations of the group A_4.

18.* Prove that the commutator subgroup of $\mathrm{GL}_n(\mathbf{R})$ is equal to $\mathrm{SL}_n(\mathbf{R})$.

4. Properties of Irreducible Complex Representations

In this section we consider only finite-dimensional representations, except for 4.1.

4.1. Morphisms. The notion of an isomorphism of linear representations was defined in the Introduction. In group theory it is well known that, in addition to isomorphisms, homomorphisms also play an important role. Similarly, arbitrary linear maps, and not only isomorphisms, are important in linear algebra. In the theory of linear representations one considers an analogous generalization of the notion of isomorphism.

Definition. Let $T_1: G \to \mathrm{GL}(V_1)$ and $T_2: G \to \mathrm{GL}(V_2)$ be linear representations of the group G. A MORPHISM OF T_1 INTO T_2 is an arbitrary linear map $\sigma: V_1 \to V_2$ satisfying the condition

(1) $$\sigma T_1(g) = T_2(g)\sigma \qquad \text{for all } g \in G.$$

[Translator's note: Such a σ is also referred to as an INTERTWINING OPERATOR, and one says that σ INTERTWINES T_1 and T_2.]

Example. Let $V = U \oplus W$ be a decomposition of the representation space of T into a direct sum of invariant subspaces. Then the projection onto U parallel to W is a morphism of T into the representation T_U.

It follows from (1) that *the kernel* $\operatorname{Ker}\sigma$ *of the morphism* σ *is a subspace invariant under* T_1, *while its image* $\operatorname{Im}\sigma = \sigma(V_1)$ *is a subspace invariant under* T_2.

If the representations T_1 and T_2 are irreducible, only two cases are possible:

1) $\operatorname{Ker}\sigma = 0, \quad \operatorname{Im}\sigma = V_2$

or

2) $\operatorname{Ker}\sigma = V_1, \quad \operatorname{Im}\sigma = 0.$

In the first case σ is an isomorphism, while in the second it is the null map. We have thus proved

Theorem 1. *Every morphism of irreducible representations is either an isomorphism or the null map.* □

In spite of its extreme simplicity, Theorem 1 has important applications. In particular, with its help one can establish the following result, which will be used in what follows.

Theorem 2. *Suppose that the space* V *of the representation* T *splits into a direct sum of minimal invariant subspaces* $V_1, \dots, V_m$ *such that the representations* $T_i = T_{V_i}$ *are pairwise nonisomorphic. Then every invariant subspace* $U \subset V$ *is the sum of a certain number of subspaces* V_i. (Cf. Remark 2 in 1.5.)

PROOF. Representation T_U is completely reducible, being a subrepresentation of the completely reducible representation T. Consequently, U is a sum of minimal invariant subspaces. It suffices to consider the case where U is itself minimal. Suppose this is the case. Let p_i denote the projection of U onto V_i. It is a morphism of the irreducible representation T_U into T_i. It follows from the assumptions of the theorem that T_U can be isomorphic only to one of the representations T_i, say, to T_1. Then p_1 is an isomorphism, whereas $p_2 = \dots = p_m = 0$. This means that $U = V_1$, which completes the proof of the theorem. □

4.2. The Schur Lemma. A morphism of a linear representation T into itself is called an ENDOMORPHISM OF T. In other words, an endomorphism of the representation T of the group G is a linear operator which commutes with all the operators $T(g)$, $g \in G$. An example is the identity operator ε.

Theorem 3 (Schur's Lemma). *Every endomorphism of an irreducible complex representation T is scalar, i.e., it has the form $c\varepsilon$, with $c \in \mathbf{C}$.*

PROOF. Let σ be an endomorphism of T:

(2) $\sigma T(g) = T(g)\sigma \quad \text{for all } g \in G.$

Let c be any of the eigenvalues of the operator σ. Then it follows from (2) that

$$(\sigma - c\varepsilon)T(g) = T(g)(\sigma - c\varepsilon) \qquad \text{for all } g \in G,$$

and so $\sigma - c\varepsilon$, too, is an endomorphism of T. Since $\det(\sigma - c\varepsilon) = 0$, Theorem 1 yields $\sigma - c\varepsilon = 0$, i.e., $\sigma = c\varepsilon$. □

Corollary. *Let T_1 and T_2 be isomorphic irreducible complex linear representations of the group G. Let σ be a fixed isomorphism of T_1 onto T_2. Then every morphism of T_1 into T_2 has the form $c\sigma$, where $c \in \mathbf{C}$.*

PROOF. Let τ be a morphism of T_1 into T_2. Then $\sigma^{-1}\tau$ is an endomorphism of T_1. By Schur's Lemma, $\sigma^{-1}\tau = c\varepsilon$, with $c \in \mathbf{C}$, and so $\tau = c\sigma$. □

Schur's Lemma permits us to describe the invariant subspaces of a completely reducible complex representation in the situation opposite to the one considered in Theorem 2, namely when all irreducible components are mutually isomorphic.

Theorem 4. *Let T be an irreducible complex representation of the group G in the space V, and let I be the trivial representation of G in the space U. Then every minimal subspace $W \subset V \otimes U$ invariant under the representation TI has the form $V \otimes u_0$, where $u_0 \in U$.*

PROOF. TI is isomorphic to the sum of a certain number of copies of the representation T (see Example 1, 3.3). By Theorem 3 of 3.2, $(TI)_W \simeq T$. Let σ be an arbitrary isomorphism of the representations $(TI)_W$ and T.

Pick a basis $(f_1, \dots, f_m)$ of U. Every element of $V \otimes U$ can be uniquely written as $x_1 \otimes f_1 + \dots + x_m \otimes f_m$, where $x_i \in V$. In particular, for every $w \in W$,

$$w = \sigma_1(w) \otimes f_1 + \dots + \sigma_m(w) \otimes f_m.$$

It is clear that the vectors $\sigma_i(w)$ depend linearly on w, and that $\sigma_i((TI)(g)w) = T(g)\sigma_i(w)$ for all $g \in G$. Hence, σ_i is a morphism of the representation $(TI)_W$ into T. By the Corollary to Theorem 3, $\sigma_i = c_i\sigma$, with $c_i \in \mathbf{C}$. Consequently,

$$w = \sigma(w) \otimes (c_1 f_1 + \dots + c_m f_m)$$

for all $w \in V$, and so indeed $W = V \otimes u_0$, where $u_0 = c_1 f_1 + \dots + c_m f_m$. □

4.3. Irreducible Representations of Abelian Groups. One of the basic facts established in linear algebra is that every linear operator in a complex vector space possesses a one-dimensional invariant subspace. This implies that every irreducible complex linear representation of a cyclic group is one-dimensional. The next result generalizes this assertion.

Theorem 5. *Every irreducible complex linear representation of an abelian group is one-dimensional.*

PROOF. Let G be an abelian group. Let T be an irreducible complex representation of G. For $g_0, g \in G$ we have

$$T(g_0)T(g) = T(g_0 g) = T(g g_0) = T(g)T(g_0).$$

This means that $T(g_0)$ is an endomorphism of T. By Schur's Lemma, $T(g_0)$ is a scalar operator. Since this holds true for every $g_0 \in G$, it follows that any subspace is invariant under T. This forces T to be one-dimensional. □

Corollary. *Every complex linear representation of an abelian group possesses a one-dimensional invariant subspace.*

PROOF. In fact, every minimal invariant subspace is, by Theorem 5, one-dimensional. □

4.4. Tensor Products of Irreducible Representations.

Theorem 6. *The tensor product of two irreducible complex representations $T: G \to \mathrm{GL}(V)$ and $S: H \to \mathrm{GL}(U)$ of the groups G and H is an irreducible representation of the group $G \times H$.*
(For the definition of the tensor product of two representations, see 3.4.)

PROOF. The tensor product $T \otimes S$ is a representation of $G \times H$ in the space $V \otimes U$. Let $W \subset V \otimes U$ be a nonnull invariant subspace. We claim that $W = V \otimes U$.

It is obvious that W is invariant under the representation $TI = (T \otimes S) \circ i_1$ of G (see 3.4). By Theorem 4, W contains a subspace of the form $V \otimes u_0$, where $u_0 \in U$, $u_0 \neq 0$.

Now for each $x \in V$ consider the subspace

$$U(x) = \{u \in U \mid x \otimes u \in W\} \subset U.$$

It is H-invariant. In fact, if $x \otimes u \in W$, then also

$$x \otimes S(h)u = (T \otimes S)(e, h)(x \otimes u) \in W.$$

Moreover, $U(x) \ni u_0$. It follows from the irreducibility of the representation S that $U(x) = U$. This means that $x \otimes u \in W$ for all $x \in V$ and $u \in U$, and so $W = V \otimes U$, as claimed. □

One can also prove the following converse of Theorem 4: *every irreducible complex linear representation of $G \times H$ is isomorpic to the tensor product of two irreducible representations of G and H.*

4.5 Spaces of Matrix Elements. Let $T: G \to \mathrm{GL}(V)$ be a complex linear representation. Let $(e) = (e_1, \dots, e_n)$ be a basis of V. We put

$$T_{(e)}(g) = [T_{ij}(g)].$$

Definition. The functions $T_{ij} \in \mathbf{C}[G]$ are called the MATRIX ELEMENTS (or MATRIX COORDINATE FUNCTIONS) OF THE REPRESENTATION T relative to the basis (e).

Any linear combination of matrix elements

$$f = \sum_{i,j} c_{ij} T_{ij} \in \mathbf{C}[G] \qquad (c_{ij} \in \mathbf{C})$$

can be expressed invariantly (without using coordinates) as

$$(3) \qquad f(g) = \operatorname{tr} \xi T(g)$$

upon denoting by ξ the linear operator given in the basis (e) by the matrix $[c_{ji}]$. It follows from this invariant expression that *the linear span of the matrix elements does not depend on the choice of a basis.*

Definition. The SPACE OF MATRIX ELEMENTS OF THE REPRESENTATION T, denoted by $\mathrm{M}(T)$, is the linear span of the matrix elements of T (relative to some basis).

We emphasize that $\mathrm{M}(T)$ is a subspace of the space $\mathbf{C}[G]$ of all $\mathbf{C}$-valued functions on the group G.

We mention *two simple properties.*

1) *If* $T_1 \simeq T_2$, *then* $\mathrm{M}(T_1) = \mathrm{M}(T_2)$. In fact, in compatible bases the representations T_1 and T_2 are given by identical matrices.

2) *If* $T = T_1 + \ldots + T_m$, *then*

$$\mathrm{M}(T) = \mathrm{M}(T_1) + \ldots + \mathrm{M}(T_m).$$

In fact, in a suitable basis the operators $T(g)$, for $g \in G$, are given by block-diagonal matrices, the diagonal blocks of which are the matrices of the operators $T_1(g), \dots, T_m(g)$ (see 3.2).

The reason for the interest attached to *the spaces of matrix elements* of various linear representations of the group G is that *they are invariant under left and right translations.*

Specifically, let f be the function given by formula (3). Then

$$f(g_2^{-1}gg_1) = \operatorname{tr} \xi\, T(g_2^{-1}gg_1) = \operatorname{tr} \xi\, T(g_2)^{-1}T(g)T(g_1) = \operatorname{tr}(T(g_1)\xi T(g_2)^{-1})T(g) = \operatorname{tr} \eta\, T(g), \tag{4}$$

where $\eta = T(g_1)\xi T(g_2)^{-1} = (T \otimes T')(g_1, g_2)\xi$ (see the Example in 3.4).

The result obtained can be interpreted as follows. Consider the map

$$\mu \colon \mathrm{L}(V) \to \mathbf{C}[G]$$

which takes each operator $\xi \in \mathrm{L}(V)$ into the function $f \in \mathbf{C}[G]$ given by (3), i.e.,

$$\mu(\xi)(g) = \operatorname{tr} \xi\, T(g) \qquad (\xi \in \mathrm{L}(V)). \tag{5}$$

Next, consider the linear representation Reg of the group $G \times G$ in $\mathbf{C}[G]$ defined by the rule

$$(\mathrm{Reg}(g_1, g_2)f)(g) = f(g_2^{-1}gg_1). \tag{6}$$

Reg combines the left and right regular representations of G.

Definition. The linear representation Reg of $G \times G$ in $\mathbf{C}[G]$ given by formula (6) is called the (TWO-SIDED) REGULAR REPRESENTATION.

Formula (4) says that *μ is a morphism of the representation $T \otimes T'$ into* Reg. The image of μ is precisely the space $\mathrm{M}(T)$ of matrix elements of T.

4.6. If T is an irreducible complex representation, then, by Theorem 4, $T \otimes T'$ is also irreducible, and so $\operatorname{Ker} \mu = 0$. We have thus proved

Theorem 7. *Let $T \colon G \to \mathrm{GL}(V)$ be an irreducible complex linear representation. Then the map μ defined by formula (5) is an isomorphism of the representations $T \otimes T'$ and $\mathrm{Reg}_{\mathrm{M}(T)}$.* □

Corollary 1. $\dim \mathrm{M}(T) = n^2$, *where $n = \dim V$.* □

Let I be the trivial representation G in a space V. Setting $g_1 = e$ or $g_2 = e$ in (4) we obtain

Corollary 2. *The map μ establishes an isomorphism of the representations IT' and $L_{\mathrm{M}(T)}$, as well as of the representations TI' and $R_{\mathrm{M}(T)}$.* □

Corollary 3. $L_{\mathrm{M}(T)} \simeq nT'$ *and* $R_{\mathrm{M}(T)} \simeq nT$. (See Example 1, 3.3.) □

Corollary 4. *Let T_1 and T_2 be nonisomorphic irreducible complex representations of the group G. Then the representations $\mathrm{Reg}_{\mathrm{M}(T_1)}$ and $\mathrm{Reg}_{\mathrm{M}(T_2)}$ of $G \times G$ are not isomorphic.*

Proof. Suppose $\mathrm{Reg}_{\mathrm{M}(T_1)}$ and $\mathrm{Reg}_{\mathrm{M}(T_2)}$ are isomorphic. Then so are their restrictions to the subgroup $G \times \{e\}$, i.e., the representations $R_{\mathrm{M}(T_1)}$ and $R_{\mathrm{M}(T_2)}$ of G. However, by Corollary 3,

$$R_{\mathrm{M}(T_1)} \simeq n_1 T_1 \quad \text{and} \quad R_{\mathrm{M}(T_2)} \simeq n_2 T_2$$

(with $n_1 = \dim T_1$ and $n_2 = \dim T_2$), and hence $R_{\mathrm{M}(T_1)} \not\simeq R_{\mathrm{M}(T_2)}$, a contradiction. □

In view of Corollary 3 of 3.2, Corollary 4 implies

Corollary 5. *Let $T_1, \dots, T_q$ be pairwise nonisomorphic irreducible complex representations of the group G. Then the subspaces $\mathrm{M}(T_1), \dots, \mathrm{M}(T_q)$ of $\mathbf{C}[G]$ are linearly independent.* □

Next we find the *explicit form of the decomposition of the space* $\mathrm{M}(T)$ *into a direct sum of minimal left-invariant subspaces.*

Let $(e) = (e_1, \dots, e_n)$ be a basis of V, and $(\varepsilon) = (\varepsilon_1, \dots, \varepsilon_n)$ be the dual basis of V'. Relative to (e), the linear operator $e_j \otimes \varepsilon_i$ is given by the matrix E_{ji} whose only nonzero entry, equal to one, is in the (j, i) site. Consequently,

$$\mu(e_j \otimes \varepsilon_i) = T_{ij}. \tag{7}$$

Proceeding from the decomposition

$$V \otimes V' = \sum (e_j \otimes V')$$

of $V \otimes V'$ into a direct sum of minimal IT'-invariant subspaces (see Example 1 of 3.3) we obtain, using the isomorphism μ, the sought-for decomposition of the space $\mathrm{M}(T)$:

$$\mathrm{M}(T) = \sum \mu(e_j \otimes V').$$

A basis of the j-th component of this decomposition is provided by the entries of the j-th column of the matrix $[T_{ij}]$.

The decomposition of $\mathrm{M}(T)$ into a direct sum of minimal right-invariant subspaces is obtained in a similar manner. A basis of the i-th component of this second decomposition is provided by the entries of the i-th row of the matrix $[T_{ij}]$.

Example. Let G be a cyclic group of order m with generator a. Consider its one-dimensional representations

$$T_k(a^x) = \omega^{kx} \qquad (k = 0, 1, \dots, m-1),$$

where $\omega = e^{\frac{2\pi i}{m}}$. They are obviously pairwise nonisomorphic. Each space $\mathrm{M}(T_k)$ is one-dimensional: it is spanned by the function T_k. By Corollary 5, the functions $T_0, T_1, \dots, T_{m-1}$ are linearly independent. This can also be verified directly: the matrix constructed from the values of these functions has the form

$$\begin{pmatrix} 1 & 1 & 1 & \cdots & 1 \\ 1 & \omega & \omega^2 & \cdots & \omega^{m-1} \\ 1 & \omega^2 & \omega^4 & \cdots & \omega^{2(m-1)} \\ \cdots & \cdots & \cdots & \cdots & \cdots \\ 1 & \omega^{m-1} & \omega^{2(m-1)} & \cdots & \omega^{(m-1)^2} \end{pmatrix},$$

and its determinant (a Vandermonde determinant) is different from zero.

4.7. Uniqueness of the Invariant Inner Product. As we saw in Section 2, introducing an invariant inner product in the representation space can be a very useful step. There arises naturally the problem of describing all such inner products. (By an inner product in a complex vector space we shall mean an arbitrary positive definite Hermitian sesquilinear function.)

To study this problem we need the following

Lemma. *Let f and f_0 be two inner products in the complex vector space V. Then there exists a linear operator σ such that*

$$(8) \qquad f(x, y) = f_0(\sigma x, y)$$

for all $x, y \in V$.

PROOF. Both sides of equation (8) are linear in x and anti-linear in y. Hence it suffices to check that (8) holds for vectors forming a basis. Let $(e) = (e_1, \dots, e_n)$ be a basis of V orthonormal with respect to the inner product f_0.

Let σ denote the linear operator given in this basis by the matrix $[f(e_j, e_i)]$. Then

$$f_0(\sigma e_i, e_j) = f_0(\sum_k f(e_i, e_k)e_k, e_j) = f(e_i, e_j),$$

and so (8) holds with the indicated σ. □

Theorem 8. *Let $T: G \to \mathrm{GL}(V)$ be an irreducible unitary representation. Then the T-invariant inner product in V is unique up to a constant factor.*

PROOF. Let f_0 and f be two invariant inner products in V. Let σ be a linear operator satisfying condition (8). We prove that σ is an endomorphism of the representation T.

For arbitrary $g \in G$ and $x, y \in V$ we have

$$\begin{aligned} f_0(T(g)^{-1}\sigma T(g)x, y) &= f_0(\sigma T(g)x, T(g)y) \\ &= f(T(g)x, T(g)y) \\ &= f(x, y) = f_0(\sigma x, y); \end{aligned} \tag{9}$$

here we used the invariance of f and f_0 under $T(g)$. Therefore, $T(g)^{-1}\sigma T(g) = \sigma$, and so $\sigma T(g) = T(g)\sigma$.

Now, by Schur's Lemma, $\sigma = c\varepsilon$ for some $c \in \mathbf{C}$. But then $f = cf_0$, as we needed to show. □

Theorem 9. *Let $T: G \to \mathrm{GL}(V)$ be a unitary representation. Let $U, W \subset V$ be minimal invariant subspaces such that $T_U \not\simeq T_W$. Then U and W are orthogonal with respect to any invariant inner product in V.*

PROOF. Fix an invariant inner-product in V and denote the corresponding orthogonal projection of the subspace W onto U by p. It is easy to see that p is a morphism of the representation T_W into T_U. By Theorem 1, $p = 0$, which means precisely that W is orthogonal to U. □

Questions and Exercises

1. Prove that the image of an invariant subspace under a morphism of representations is an invariant subspace.

2.* Let G be a doubly-transitive group of permutations, i.e., a subgroup of the symmetric group S_n with the following property: for any i, j, k, ℓ such that $i \neq j$ and $k \neq \ell$ there exists a permutation $\sigma \in G$ such that $\sigma(i) = k$ and $\sigma(j) = \ell$. Let M be a monomial representation of S_n (see Example 5 of 1.3). Prove that every endomorphism of the representation $M|_G$ has the form $a\varepsilon + b\eta$, where $a, b \in \mathbf{C}$ and $\eta(e_i) = e_1 + \ldots + e_n$ for all i.

3. Let $T_1, \ldots, T_q$ be pairwise nonisomorphic complex linear representations of the group G. Prove that the set of all morphisms of the representation $\sum k_i T_i$ into the representation $\sum \ell_i T_i$ is a vector space of dimension $\sum k_i \ell_i$.

4.* Using Problems 2 and 3, prove that if G is a doubly-transitive group of permutations, then the representation $M_0|_G$ (see Example 5 of 1.3) is irreducible.

5. Find all automorphisms of the representation of $\mathbf{R}$ by rotations in the Euclidean plane.

6. Let T be an arbitrary complex representation of the abelian group G. Show that in the representation space of T there is a basis relative to which all operators $T(g)$, for $g \in G$, are given by triangular matrices.

7. Prove that every irreducible real representation of an abelian group is one- or two-dimensional.

8.* Let G be a finite group and R the right regular representation of G (see 0.9). Give a direct proof of the fact that the dimension of the space of all morphisms of R into an irreducible representation T is equal to $\dim T$. Applying Problem 3, deduce from this that $R \simeq \sum_{i=1}^{q} (\dim T_i) T_i$, where $T_1, \ldots, T_q$ is a complete list of irreducible complex representations of G.

9. Under the assumptions of Theorem 4, prove that every G-invariant subspace of $V \otimes U$ has the form $V \otimes U_0$, where U_0 is a subspace of U.

10. Prove that the matrix elements of an irreducible complex representation are linearly independent. Is this assertion true for real representations?

11. Let $T: G \to \mathrm{GL}(V)$ be an irreducible complex representation. Prove that the linear span of the set $\{T(g) \mid g \in G\} \subset \mathrm{L}(V)$ equals $\mathrm{L}(V)$ (Burnside's Theorem).

12. Prove that every irreducible representation of the group G over an arbitrary field is isomorphic to a subrepresentation of the right regular representation of G.

13. The same for the left regular representation.

14.* Prove Corollaries 4 and 5 of Theorem 7 for linear representations over an arbitrary field.

15.* Let $T: G \to \mathrm{GL}(V)$ be an irreducible orthogonal representation. Prove that the invariant inner product in V is unique up to a (positive) constant factor.

II. Representations of Finite Groups

In this chapter we are concerned only with finite-dimensional representations, mainly complex ones. Recall that in Section 2 we have shown that every complex linear representation of a finite group is unitary, and hence completely reducible.

The discussion in 5.3 and 5.5 is valid for arbitrary, and not only finite groups.

5. Decomposition of the Regular Representation

5.1. Let G be a *finite* group, and let $\mathbf{C}[G]$ be the space of $\mathbf{C}$-valued functions on G. Then

$$\dim \mathbf{C}[G] = |G| < \infty.$$

All our results concerning linear representations of finite groups will rely on the study of the *regular representations* Reg, L, and R. We remind the reader that

$$(\mathrm{Reg}(g_1, g_2)f)(g) = f(g_2^{-1} g g_1),$$
$$(L(g_2)f)(g) = f(g_2^{-1} g),$$

and

$$(R(g_1)f)(g) = f(g g_1)$$

for every function $f \in \mathbf{C}[G]$.

By Corollary 3 to Theorem 7 of 4.6, every irreducible complex representation of G is isomorphic to a subrepresentation of R. Since R is finite-dimensional, this yields

Theorem 1. *A finite group has only finitely many irreducible complex representations, up to isomorphism.* □

E.B. Vinberg, *Linear Representations of Groups*, Modern Birkhäuser Classics,
DOI 10.1007/978-3-0348-0063-1_3, © Birkhäuser Verlag 1989

5.2. To formulate the main results of this section, we fix the following notations:

G is a finite group;

$T_k: G \to \mathrm{GL}(V_k)$, $k = 1, \dots, q$, is a full list of its irreducible complex representations;

$n_k = \dim V_k$;

I_k is the trivial representation of G in the space V_k;

$\mu_k: \mathrm{L}(V_k) \to \mathbf{C}[G]$ is the linear map constructed for T_k following the rule given in 4.5: $\mu_k(\xi)(g) = \operatorname{tr} \xi T_k(g)$, for any $\xi \in \mathrm{L}(V_k)$;

$\mathrm{M}(T_k) = \operatorname{Im} \mu_k$ is the space of matrix elements of the representation T_k (see 4.5).

The main theorem of this section is

Theorem 2. $\mathbf{C}[G] = \mathrm{M}(T_1) \oplus \dots \oplus \mathrm{M}(T_q)$.

PROOF. By Corollary 5 to Theorem 7 of 4.6, the subspaces $\mathrm{M}(T_1), \dots, \mathrm{M}(T_q)$ are linearly independent. We show that their sum is equal to $\mathbf{C}[G]$.

Let $(f_1, \dots, f_N)$ be a basis of $\mathbf{C}[G]$. Let R_{ij} denote the matrix elements of the right regular representation relative to this basis. Then

$$f_j(g) = (R(g)f_j)(e) = (\sum_i R_{ij}(g) f_i)(e) = \sum_i f_i(e) R_{ij}(g),$$

i.e., $f_j = \sum_i f_i(e) R_{ij}$. Thus, the basis functions f_j and, generally, all functions on G belong to the space $\mathrm{M}(R)$ of matrix elements of the representation R. Since $T_1, \dots, T_q$ is the set of all irreducible complex representations of the group G,

$$R \simeq \sum m_k T_k,$$

and so

$$\mathrm{M}(R) \subset \mathrm{M}(T_1) + \dots + \mathrm{M}(T_q)$$

(see 4.5). Consequently,

$$\mathbf{C}[G] = \mathrm{M}(T_1) + \dots + \mathrm{M}(T_q),$$

as claimed. □

Corollary 1. $n_1^2 + \dots + n_q^2 = |G|$. □

Theorem 2, in conjunction with Corollary 3 to Theorem 7 of 4.6, yields

Corollary 2. $L \simeq R \simeq \sum n_k T_k$. □

(From the foregoing analysis it follows immediately that $L \simeq \sum n_k T_k'$; however, since a dual representation has the same dimension as the original one, $\sum n_k T_k' \simeq \sum n_k T_k$.)

Examples.

1. If the group G is abelian, then all its irreducible complex representations are one-dimensional (Theorem 5, 4.6). Hence the number of such representations is equal to the order of G.

2. The group $G = S_3$ clearly possesses the following irreducible complex representations: the two one-dimensional ones described in 3.9, and the two-dimensional one of Example 2, 3.7. Since

$$1^2 + 1^2 + 2^2 = 6 = |S_3|,$$

S_3 has no other irreducible complex representations. In particular, the representation M_0 constructed in Example 5 of 1.3 is isomorphic to the two-dimensional representation indicated above.

5.3. Characters of Linear Representations. Let $T: G \to \mathrm{GL}(V)$ be a linear representation.

Definition. The function

$$\chi_T: g \mapsto \operatorname{tr} T(g)$$

is called the CHARACTER OF THE REPRESENTATION T.

In other words, χ_T is *the sum of the diagonal matrix elements of* T. Notice that $\chi_T \in \mathrm{M}(T)$.

Examples.

1. The character of a one-dimensional representation coincides with the representation itself.

2. The character of a trivial representation is a constant, equal to the dimension of the representation.

3. The character of the representation of $\mathbf{R}$ by rotations in the plane is the function $t \mapsto 2\cos t$.

4. The character of a monomial representation M of S_n (Example 5, 1.3) is the function which assigns to each permutation the number of its fixed points.

5. The character of the two-dimensional representation T of S_3 constructed in Example 2 of 3.7, is given by the formula

$$\chi_T(\sigma) = \begin{cases} 2, & \text{if } \sigma = \varepsilon, \\ 0, & \text{if } \sigma \text{ is a transposition}, \\ -1, & \text{if } \sigma \text{ is a triple cycle}. \end{cases}$$

This follows from the fact that, in a suitable basis, the matrix of the operator $T(\sigma)$ has the form

$$\begin{pmatrix} 1 & 0 \\ 0 & 1 \end{pmatrix}, \quad \begin{pmatrix} 1 & 0 \\ 0 & -1 \end{pmatrix}, \quad \text{or} \quad \begin{pmatrix} -\frac{1}{2} & -\frac{\sqrt{3}}{2} \\ \frac{\sqrt{3}}{2} & -\frac{1}{2} \end{pmatrix},$$

depending on which of the three indicated cases σ falls into.

Since isomorphic representations are given in compatible bases by identical matrices, we see that *isomorphic representations have equal characters.*

The character of the dual representation T' satisfies the relation

$$\chi_{T'}(g) = \chi_T(g^{-1}),$$

which follows from formula (1) of 3.1.

The characters of sums and products of representations are calculated by the formulas

(1) $$\chi_{T+S} = \chi_T + \chi_S$$

and respectively

(2) $$\chi_{TS} = \chi_T \chi_S.$$

The first of these is self-evident, while the second follows from the general property $\operatorname{tr}(\alpha \otimes \beta) = \operatorname{tr} \alpha \cdot \operatorname{tr} \beta$, established in Appendix 2.

Example. In 1.3 we showed that the monomial representation M of the group S_n decomposes into the sum of the trivial one-dimensional representation and the irreducible representation M_0. Correspondingly,

$$\chi_M(\sigma) = \chi_{M_0}(\sigma) + 1$$

for all $\sigma \in S_n$. Since the character of M is already known (see Example 4 above), this formula permits us to find the character of M_0. In particular, for $n = 3$ we get

$$\chi_{M_0}(\sigma) = \begin{cases} 2, & \text{if } \sigma = \varepsilon, \\ 0, & \text{if } \sigma \text{ is a transposition}, \\ -1, & \text{if } \sigma \text{ is a triple cycle}. \end{cases}$$

The representation M_0 of S_3 is isomorphic to the representation constructed in Example 2, 3.7 (see Example 2, 5.2), and our result agrees with the one obtained above in Example 5.

The most important property of the character of a representation is that it is constant on every conjugacy class of elements of the group (see Appendix 4), i.e.,

(3) $\qquad \chi_T(gxg^{-1}) = \chi_T(x)$

for all $g, x \in G$. This follows from the facts that the matrices of the operators $T(gxg^{-1})$ and $T(x)$ are similar and similar matrices have the same trace.

Definition. A function f on the group G with the property that

$$f(gxg^{-1}) = f(x) \qquad \text{for all } g, x \in G$$

is called a CENTRAL FUNCTION.

Thus, *the character of any representation is a central function.* We could have verified this in each of the examples discussed above.

5.4. It is clear that the central functions form a subspace in the space of all functions on G. We denote it by $\mathbf{C}[G]^{\#}$.

If the group G is finite, then the dimension of $\mathbf{C}[G]^{\#}$ is equal to the number of distinct conjugacy classes of G.

Theorem 3. *The characters of the irreducible complex representations of a finite group G constitute a basis of the space $\mathbf{C}[G]^{\#}$.*

PROOF. Preserving the notations of 5.2, we let χ_k denote the character of the representation T_k.

By Theorem 2, each function $f \in \mathbf{C}[G]$ is uniquely expressible as

(4) $\qquad f = f_1 + \ldots + f_q, \qquad f_k \in \mathrm{M}(T_k).$

According to the definition, f is central if

(5) $\qquad \mathrm{Reg}(g, g)f = f$

for all $g \in G$. Since the subspaces $\mathrm{M}(T_k)$ are invariant under Reg, (5) implies that

$$\mathrm{Reg}(g, g)f_k = f_k$$

for all k and all $g \in G$, i.e., each f_k is a central function. Hence, in order to prove the theorem it suffices to show that every central function belonging to $\mathrm{M}(T_k)$ is proportional to χ_k. This is the objective of the following lemma.

Lemma. *Let $T: G \to \mathrm{GL}(V)$ be an irreducible complex representation of the group G, and let $\chi = \chi_T$ be its character. Then every central function f belonging to $\mathrm{M}(T)$ is proportional to χ_T.*

PROOF. We use the isomorphism μ of the representations $T \otimes T'$ and $\mathrm{Reg}_{\mathrm{M}(T)}$ (Theorem 7, 4.6). Let $f = \mu(\xi)$, with $\xi \in \mathrm{L}(V)$. Since μ is an isomorphism, the fact that f is central, expressed by equation (5), translates into the equality

$$T(g)\xi T(g)^{-1} = \xi$$

for all $g \in G$, i.e., into ξ being an endomorphism of the representation T. By Schur's Lemma, $\xi = c\varepsilon$, with $c \in \mathbf{C}$. Now observe that

$$(\mu(\varepsilon))(g) = \operatorname{tr} \varepsilon T(g) = \operatorname{tr} T(g) = \chi(g),$$

i.e., $\mu(\varepsilon) = \chi$. Consequently, $\mu(\xi) = c\chi$, as needed. □

Corollary 1. *The number of irreducible complex representations of a finite group G is equal to the number of its distinct conjugacy classes.* □

Corollary 2. *An irreducible complex representation of a finite group G is uniquely determined, up to an isomorphism, by its character.*

PROOF. Every complex representation T of the group G has the form

$$T = \sum_k m_k T_k,$$

and then $\chi_T = \sum_k m_k \chi_k$. Since the characters χ_k are linearly independent (which, in fact, is the only part of Theorem 3 that we need), the coefficients m_k are uniquely determined by χ_T, and hence so is the representation T. □

Examples.

1. In an *abelian group* every conjugacy class consists of a single element. Accordingly, the number of irreducible complex representations of a finite abelian group is equal to its order (another proof of this fact was given in Example 1 of 5.2).

2. In the group S_4 there are five conjugacy classes, with representatives $\varepsilon, (12), (12)(34), (123)$, and (1234). We already know the following four irreducible complex representations of S_4: two one-dimensional representations (see the Example in 3.9), the two-dimensional one (Example 2, 3.8), and the three-dimensional representation M_0 (Example 5, 1.3). The sum of the

squares of their dimensions is $1^2 + 1^2 + 2^2 + 3^2 = 15$. Hence S_4 possesses one more irreducible complex representation, of dimension $\sqrt{24-15} = 3$.

The representation missing in our list can be described as a product $M_0\Pi$, where Π is the nontrivial one-dimensional representation. In fact, $M_0\Pi$ is irreducible. Also, it is not isomorphic to M_0, since $\det(M_0\Pi)(g) = 1$ for all $g \in S_4$, whereas the representation M_0 does not have this property. It is readily established that $(M_0\Pi)(S_4)$ is the group of rotations of a cube, whereas $M_0(S_4)$ is the full group of symmetries of a regular tetrahedron.

5.5. In applications of group theory the regular representation per se is rarely encountered. However, one is often led to considering its subrepresentations connected with the so-called *transitive actions* of the given group.

Definition. An action $s: G \to S(X)$ is said to be TRANSITIVE if for any $x, y \in X$ there is a $g \in G$ such that $s(g)x = y$.

Examples.

1. The natural action of S_n on the set $\{1, 2, \ldots, n\}$.

2. The action of the orthogonal group on the unit sphere.

An important and, as we shall see presently, universal example of a transitive action is l^H, *the action of the group G on the set G/H of left cosets of the subgroup H in G* defined by the rule

$$l^H(g)(uH) = guH.$$

For $H = \{e\}$, $l^H = l$, the action of G on itself by left translations.

Now let $s: G \to S(X)$ be an arbitrary transitive action. Pick an arbitrary point $o \in X$ and consider the map

$$p: G \to X, \qquad g \mapsto go.$$

It follows from the transitivity of s that $p(G) = X$. Let H denote the ISOTROPY SUBGROUP of the point o:

$$H = \{g \in G \mid go = o\}.$$

The preimage of any point $x = go \in X$ under p is precisely the left coset gH. Hence there is a bijective map

$$\bar{p}: G/H \to X$$

sending every coset gH into the point go (which does not depend on the choice of a representative in the coset). It commutes with the actions of G:

$$\bar{p}l^H(g) = s(g)\bar{p} \qquad \text{for all } g \in G;$$

in fact, for every coset uH we have

$$\bar{p}l^H(g)(uH) = \bar{p}(guH) = guo$$

and

$$s(g)\bar{p}(uH) = s(g)(uo) = guo.$$

Thus, *every transitive action is isomorphic to an action of the form* l^H. Accordingly, the corresponding linear representation in the space of functions on X is isomorphic to the linear representation $l_*^H = L^H$ in the space of functions on G/H.

Moreover, functions on G/H may be regarded as functions on G that are constant on the cosets of H in G. Under this convention, L^H turns into the subrepresentation of L corresponding to the invariant subspace

$$\begin{aligned} \mathbf{C}[G]^H &= \{ f \in \mathbf{C}[G] \mid f(gh) = f(g) \text{ for all } g \in G, h \in H \} \\ &= \{ f \in \mathbf{C}[G] \mid R(h)f = f \text{ for all } h \in H \} \end{aligned} \tag{6}$$

consisting of the functions constant on the cosets of H in G.

The next subsection is devoted to the study of representations of the form L^H for finite groups.

5.6. Theorem 4. *Let H be a subgroup of the finite group G. Then*

$$\mathbf{C}[G]^H = \sum \mu_k(V_k^H \otimes V_k'), \tag{7}$$

where

$$V_k^H = \{ x \in V_k \mid T_k(h)x = x \text{ for all } h \in H \}. \tag{8}$$

PROOF. Since $\mathbf{C}[G] = \sum_k \mathrm{M}(T_k)$ and the subspaces $\mathrm{M}(T_k)$ are invariant under right translations,

$$\mathbf{C}[G]^H = \sum_k \mathrm{M}(T_k)^H$$

where

$$\mathrm{M}(T_k)^H = \{ f \in \mathrm{M}(T_k) \mid R(h)f = f \text{ for all } h \in H \}.$$

By Theorem 7 of 4.6, the representation $R_{\mathrm{M}(T_k)}$ is isomorphic to $T_k I_k'$, and the isomorphism is realized by the map μ_k. Let $(\varepsilon_1, \dots, \varepsilon_n)$ be a basis of V_k'. Every function $f \in \mathrm{M}(T_k)$ is uniquely expressible as

$$f = \mu_k(\sum_i x_i \otimes \varepsilon_i), \qquad \text{with } x_i \in V_k.$$

Then

$$R(h)f = \mu_k(\sum_i T_k(h)x_i \otimes \varepsilon_i)$$

for all $h \in H$. Hence, $f \in \mathrm{M}(T_k)^H$ if and only if $x_i \in V_k^H$ for all i. We conclude that $\mathrm{M}(T_k)^H = \mu_k(V_k^H \otimes V_k')$, as needed. □

Corollary. $L^H \simeq \sum_k m_k T_k$, *where* $m_k = \dim V_k^H = \dim (V_k')^H$.

PROOF. The map μ_k is an isomorphism of the representations $L_{\mathrm{M}(T_k)}$ and $I_k T_k'$. Consequently,

$$L_{\mathrm{M}(T_k)^H} \simeq m_k T_k',$$

where $m_k = \dim V_k^H$, and so

$$L^H = \sum_k L_{\mathrm{M}(T_k)^H} \simeq \sum_k m_k T_k' = \sum_k m_k' T_k,$$

where $m_k' = \dim (V_k')^H$. It remains to show that $m_k = m_k'$, which is done by applying the following general lemma to the representation $S = T_k|_H$.

Lemma. *Let* $S\colon H \to \mathrm{GL}(V)$ *be a completely reducible linear representation of the group* H. *Then* $\dim V^H = \dim (V')^H$.

PROOF. Let $S = \sum S_i$ be the decomposition of S into a sum of irreducible representations. Then $\dim V^H$ is equal to the number of trivial representations among the S_i's. The dual representation S' admits the decomposition $S' = \sum S_i'$, and the lemma follows on observing that S_i' is trivial if and only if S_i is trivial. □

Example. Let G be the group of rotations of a cube, and let s be its natural action on the set X of faces of the cube (see Example 1, 0.9). Since s is transitive, the representation s_* is isomorphic to L^H, where H is the isotropy subgroup of an arbitrary face. It is clear that H is a cyclic group of order 4. It is known (see, for example, [8]), that $G \simeq S_4$. The isomorphism $G \to S_4$ takes H into the subgroup $< (1234) >$.

As we saw in Example 4 of 5.4, S_4 has five irreducible complex representations of dimensions $1, 1, 2, 3$, and 3. It is readily seen that the subspace of H-invariant vectors is one-dimensional for the trivial one-dimensional representation I, the two-dimensional representation, and the three-dimensional representation $M_0\Pi$, and that it is zero-dimensional for the representations Π and M_0. $M_0\Pi$ corresponds to the identity representation Id of G.

Therefore

$$(9) \qquad s_* \simeq I + P + \mathrm{Id},$$

where P denotes an irreducible two-dimensional representation. The invariant subspaces $U_1, U_2, U_3 \subset \mathbf{C}[X]$ corresponding to the decomposition (9) are described as follows:

U_1 consists of the constant functions,

U_2 consists of the "even" functions (i.e., functions taking identical values on opposite faces of the cube) with null sums of values, and

U_3 consists of the "odd" functions (i.e., functions taking opposite values on opposite faces of the cube).

5.7. Representations of the group A_5. As an illustration of the results obtained in this section we find all irreducible complex representations of the group A_5 of even permutations of degree five.

In A_5 there are five distinct conjugacy classes, with the permutations ε, (12)(34), (123), (12345), and (21345) as representatives. Accordingly, A_5 possesses five irreducible complex representations, denoted here by T_1 through T_5. Setting $n_k = \dim T_k$, we have $\sum_{k=1}^{5} n_k^2 = |A_5| = 60$.

The restriction of the representation M_0 of S_5 to A_5 is irreducible (see Exercise 11, 1.6). It is four-dimensional, and we denote it by T_4.

Also, we let T_1 denote the trivial one-dimensional representation.

Since $n_1^2 + n_4^2 = 17$, we have $n_2^2 + n_3^2 + n_5^2 = 60 - 17 = 43$, which yields $n_2 = n_3 = 3$ and $n_5 = 5$.

It is known (see, for example, [8]), that A_5 is isomorphic to the group $G(D)$ of rotations of a regular dodecahedron D. An arbitrarily fixed isomorphism $A_5 \to G(D)$ yields a three-dimensional irreducible representation T_2. Let a be the outer automorphism of A_5 given by

$$a(\sigma) = (12)\sigma(12)^{-1} \qquad (\sigma \in A_5).$$

Then $T_2 \circ a = T_3$ is a three-dimensional irreducible representation of A_5 that is not isomorphic to T_2. In fact, a maps the two conjugacy classes of elements of order five existing in A_5 into one another. The elements of one of these classes are taken by the representation T_2 into rotations through $2\pi/5$, and the elements of the second into rotations through $4\pi/5$; under the representation T_3 one has the opposite situation.

The remaining five-dimensional irreducible representation T_5 can be found upon decomposing the representation s_* associated with the transitive action s of the group $A_5 \simeq G(D)$ on the set X of faces of D. The isotropy subgroup for this action is a cyclic group of order five. Using the results of 5.6, one can show that

$$s_* \simeq L^H \simeq T_1 + T_2 + T_3 + T_5. \tag{10}$$

The representation T_5 is realized in the subspace of $\mathbf{C}[X]$ consisting of the even functions (see the Example, 5.6) with null sums of values.

Questions and Exercises

1. Construct a basis of matrix elements in the space $\mathbf{C}[S_3]$.

2. What is the value of a character at the identity element of the group?

3. Calculate the characters

a) of all irreducible complex representations of S_3;

b) of the left and right regular representations of an arbitrary finite group.

4. Find all irreducible complex representations of the group A_4 and calculate their characters.

5. Suppose that all irreducible complex representations of the group G are one-dimensional. Prove that G is abelian.

6.* Prove that every finite group of order larger than 2 has more than two irreducible complex representations.

7. Find the irreducible complex representations

a) of the dihedral group

$$D_n = < a, b \mid a^n = b^2 = 1, \quad bab^{-1} = a^{-1} >;$$

b)*of the generalized group of quaternion units

$$Q_n = < a, b \mid a^{2n} = b^4 = 1, \quad bab^{-1} = a^{-1}, \quad b^2 = a^n > .$$

Verify Corollary 1 to Theorem 2 and Corollary 1 to Theorem 3 for these examples.

8. By computing characters, find out for which values of n the representations M_0 and $M_0\Pi$ of the group S_n are not isomorphic.

9. Let H be a subgroup of the finite group G. In the notations of 5.2 and 5.6, show that

$$\sum_k m_k n_k = [G : H],$$

(where $[G : H]$ denotes the order of H in G).

10. Decompose the representation s_* into a sum of irreducible representations in the case where s is

a) the action of the group of rotations of a cube on the set of its vertices;

b) the action of the full symmetry group of a regular tetrahedron on the set of its edges.

11. Find the character of the five-dimensional irreducible complex representation of A_5.

12. Prove that a representation of a finite group is isomorphic to its dual if and only if its character takes on only real values.

6. Orthogonality Relations

Throughout this section G is a finite group. We preserve the notations of 5.2, and $\mathbf{C}[G]^\#$, χ_k of 5.4.

The space $\mathbf{C}[G]$ of functions on the group G is equipped with the *inner product* defined by the formula

$$(1) \qquad (f_1, f_2) = \frac{1}{|G|} \sum_{g \in G} f_1(g)\overline{f_2(g)}.$$

We show that $(\cdot,\cdot)$ is invariant under left and right translations, i.e., under the regular representation Reg of $G \times G$ in $\mathbf{C}[G]$.

In fact,

$$(2) \qquad (\mathrm{Reg}(g_1, g_2)f_1, \mathrm{Reg}(g_1, g_2)f_2) = \frac{1}{|G|} \sum_{g \in G} f_1(g_2^{-1}gg_1)\overline{f_2(g_2^{-1}gg_1)}$$

for every $g_1, g_2 \in G$. Since the equation $g_2^{-1}xg_1 = g$ has a unique solution for every $g \in G$, the sum in (2) differs from that in (1) only in the order of the terms. Consequently,

$$(\mathrm{Reg}(g_1, g_2)f_1, \mathrm{Reg}(g_1, g_2)f_2) = (f_1, f_2),$$

as claimed.

The main theorem of this section describes the inner products of the matrix elements of the irreducible representations of G.

Let $T_{k,ij}$ denote the matrix elements of the representation T_k relative to an orthonormal basis of the space V_k (it is assumed that the inner product in V_k is T_k-invariant).

Theorem 1. *The matrix elements* $T_{k,ij}$ *constitute an orthogonal basis in* $\mathbf{C}[G]$. *Moreover*

$$(T_{k,ij}, T_{k,ij}) = n_k^{-1}.$$

PROOF. We first show that for $k \neq \ell$ the matrix elements of the representations T_k and T_ℓ are orthogonal. To this end we apply Theorem 9 of 4.7 to the representation Reg. The representations $\mathrm{Reg}_{\mathrm{M}(T_k)}$ and $\mathrm{Reg}_{\mathrm{M}(T_\ell)}$ are not isomorphic (Corollary 4 to Theorem 7, 4.6). Hence, the subspaces $\mathrm{M}(T_k)$ and $\mathrm{M}(T_\ell)$ are orthogonal.

To evaluate the inner products of the matrix elements of the representation T_k, we use the isomorphism

$$\mu_k \colon \mathrm{L}(V_k) \to \mathrm{M}(T_k).$$

In $\mathrm{L}(V_k)$ we define an inner product by the rule

$$(3) \qquad (\xi, \eta) = \operatorname{tr} \xi\eta^*,$$

where η^* denotes the adjoint of the operator η. This inner product is invariant under the representation $T_k \otimes T_k'$. In fact,

$$\begin{aligned}(T(g_1)\xi T(g_2)^{-1}, T(g_1)\eta T(g_2)^{-1})& \\ &= \operatorname{tr} T(g_1)\xi T(g_2)^{-1} T(g_2)^{*-1}\eta^* T(g_1)^* \\ &= \operatorname{tr} T(g_1)\xi\eta^* T(g_1)^{-1} = \operatorname{tr} \xi\eta^* = (\xi, \eta)\end{aligned}$$

for all $g_1, g_2 \in G$ (here we used the fact that the operators $T(g_1)$ and $T(g_2)$ are unitary, and hence their adjoints coincide with their inverses).

Let $(e) = (e_1, \dots, e_{n_k})$ be an orthonormal basis of V_k, and let $(\varepsilon) = (\varepsilon_1, \dots, \varepsilon_{n_k})$ be the dual basis of V_k'. We claim that the operators $e_i \otimes \varepsilon_j$ constitute an orthonormal basis in the space $\mathrm{L}(V_k) = V_k \otimes V_k'$. In fact, the matrix of $e_i \otimes \varepsilon_j$ is E_{ij}, and the matrix of the adjoint operator is $\bar{E}_{ij}' = E_{ji}$. Hence

$$(e_i \otimes \varepsilon_j, e_s \otimes \varepsilon_t) = \operatorname{tr} E_{ij} E_{ts} = \begin{cases} 1, & \text{if } i = s,\ j = t, \\ 0, & \text{otherwise.} \end{cases}$$

We carry the inner product (3) over to the space $\mathrm{M}(T_k)$ using the mapping μ_k. Since μ_k is an isomorphism of the representations $T_k \otimes T_k'$ and $\mathrm{Reg}_{\mathrm{M}(T_k)}$, we obtain in this manner a Reg-invariant inner product in $\mathrm{M}(T_k)$. By Theorem 8 of 4.7, the latter must be proportional to the inner product (1). This means that for every $\xi, \eta \in \mathrm{L}(V_k)$ we have

$$(4) \qquad (\mu_k(\xi), \mu_k(\eta)) = c_k \operatorname{tr} \xi\eta^*,$$

where c_k is a constant.

As $T_{k,ij} = \mu_k(e_j \otimes \varepsilon_i)$ (formula (7), 4.6), it follows from (4) that, for each fixed k, the functions $T_{k,ij}$ form an orthogonal basis of the space $\mathrm{M}(T_k)$, and that

$$(T_{k,ij}, T_{k,ij}) = c_k.$$

It remains to show that $c_k = n_k^{-1}$.

To this end, we notice that from the unitarity of the operators $T_k(g)$ it follows that

$$\sum_j T_{k,ij}(g)\overline{T_{k,ij}(g)} = 1$$

for all $g \in G$. Summing over $g \in G$ and dividing by $|G|$ we get

$$\sum_j (T_{k,ij}, T_{k,ij}) = 1,$$

i.e., $n_k c_k = 1$, as needed. □

Corollary 1. *The characters $\chi_1, \dots, \chi_q$ form an orthonormal basis in the space $\mathbf{C}[G]^\#$.*

PROOF. We have $\chi_k = \sum_i T_{k,ii}$. Consequently, $(\chi_k, \chi_\ell) \neq 0$ for $k \neq \ell$, whereas

$$(\chi_k, \chi_k) = \sum_i (T_{k,ii}, T_{k,ii}) = n_k n_k^{-1} = 1.$$

□

The main application of the orthogonality relations is

Corollary 2. *Let T be a complex linear representation of the group G. Then $T \simeq \sum m_k T_k$, where*

$$m_k = (\chi_T, \chi_k). \tag{5}$$

PROOF. If $T \simeq \sum m_k T_k$, then $\chi_T = \sum m_k T_k$ and, by Corollary 1, $(\chi_T, \chi_k) = m_k$. □

Thus, once we know the character χ_T of the representation T, we can find the decomposition of T into a sum of irreducible representations, referred to as the SPECTRUM of T. In this way we can, in particular, find the spectrum of the product of two representations; indeed, the character of the product is equal to the product of the characters of the two given representations (see 5.3).

We mention also

Corollary 3. *The complex representation T of G is irreducible if and only if*

$$(\chi_T, \chi_T) = 1.$$

PROOF. Suppose $T \simeq \sum m_k T_k$. Then $(\chi_T, \chi_T) = \sum m_k^2$. Obviously, $\sum m_k^2 = 1$ if and only if one of the numbers m_k is equal to 1 and the others to 0. □

Examples.

1. The matrix elements of the complex irreducible representations of a finite abelian group G coincide with the characters of G and hence with the representations themselves. The meaning of the orthogonality relations is that the matrix built from the values of the matrix elements, each divided by $|G|^{1/2}$, is unitary. For example, if G is a cyclic group of order m, this matrix has the form (see the Example of 4.6)

$$\frac{1}{\sqrt{m}} \begin{pmatrix} 1 & 1 & 1 & \dots & 1 \\ 1 & \omega & \omega^2 & \dots & \omega^{m-1} \\ 1 & \omega^2 & \omega^4 & \dots & \omega^{2(m-1)} \\ \dots & \dots & \dots & \dots & \dots \\ 1 & \omega^{m-1} & \omega^{2(m-1)} & \dots & \omega^{(m-1)^2} \end{pmatrix}, \quad \text{with } \omega = e^{\frac{2\pi i}{m}}.$$

2. Using the description of the irreducible complex representations of the group A_5 given in 5.7, we can write down the following table of characters of A_5:

	ε	$(12)(34)$	(123)	(12345)	(21345)
χ_1	1	1	1	1	1
χ_2	3	-1	0	$\frac{1}{2}(1+\sqrt{5})$	$\frac{1}{2}(1-\sqrt{5})$
χ_3	3	-1	0	$\frac{1}{2}(1-\sqrt{5})$	$\frac{1}{2}(1+\sqrt{5})$
χ_4	4	0	1	-1	-1
χ_5	5	1	-1	0	0
	1	15	20	12	12

Here in the first and last rows we indicate representatives of the conjugacy classes and the numbers of elements in these classes respectively; the latter are needed in order to compute the inner products according to formula (1).

The characters χ_2 and χ_3 are found from the geometrical description of the representations T_2 and T_3. For instance, the operator $T_2((12345))$ is a rotation through an angle of $2\pi/5$; relative to a suitable basis it is described by the matrix

$$\begin{pmatrix} \cos\frac{2\pi}{5} & -\sin\frac{2\pi}{5} & 0 \\ \sin\frac{2\pi}{5} & \cos\frac{2\pi}{5} & 0 \\ 0 & 0 & 1 \end{pmatrix},$$

the trace of which is $2\cos\frac{2\pi}{5} + 1 = \frac{1}{2}(1+\sqrt{5})$.

The character χ_5 can be found from formula (10) of 5.7.

The characters found above satisfy the orthogonality relations. For example,

$$(\chi_4, \chi_5) = \frac{1}{60}(4 \cdot 5 + 20 \cdot 1 \cdot (-1)) = 0.$$

The table of characters can be used to *compute spectra of products of representations*. For example, let us find the character of the representation T_2^2. By formula (2) of 5.3, it is equal to χ_2^2. We write its values in a table

	ε	(12)(34)	(123)	(12345)	(21345)
χ_2^2	9	1	0	$\frac{1}{2}(3+\sqrt{5})$	$\frac{1}{2}(3-\sqrt{5})$

We next compute the inner products (χ_2^2, χ_k):

$$\begin{aligned}
(\chi_2^2, \chi_1) &= \frac{1}{60}\left[9 + 15 \cdot 1 + 12 \cdot \frac{1}{2}(3+\sqrt{5}) + 12 \cdot \frac{1}{2}(3-\sqrt{5})\right] = 1, \\
(\chi_2^2, \chi_2) &= \frac{1}{60}\left[3 \cdot 9 + 15 \cdot (-1) + 12 \cdot (2+\sqrt{5}) + 12 \cdot (2-\sqrt{5})\right] = 1, \\
(\chi_2^2, \chi_3) &= \frac{1}{60}\left[3 \cdot 9 + 15 \cdot (-1) - 12 \cdot \frac{1}{2}(1+\sqrt{5}) - 12 \cdot \frac{1}{2}(1-\sqrt{5})\right] = 0, \\
(\chi_2^2, \chi_4) &= \frac{1}{60}\left[4 \cdot 9 - 12 \cdot \frac{1}{2}(3+\sqrt{5}) - 12 \cdot \frac{1}{2}(3-\sqrt{5})\right] = 0, \\
(\chi_2^2, \chi_5) &= \frac{1}{60}(5 \cdot 9 + 15) = 1.
\end{aligned}$$

Applying Corollary 2 of Theorem 1, we conclude that

(6) $\quad T_2^2 \simeq T_1 + T_2 + T_5.$

Questions and Exercises

1. Let χ_R denote the character of the right regular representation R. Compute the inner product (χ_R, χ_R) directly, and also by decomposing R into a sum of irreducible representations.

2. Let T be a complex linear representation of the group G in a space V. Show that

$$\dim V^G = \frac{1}{|G|} \sum_{g \in G} \chi_T(g).$$

3. Write down the table of characters of S_4 and find the spectrum of the representation M_0^2.

4. Use the theory of characters to find the spectra of the representations of S_4 (see Exercise 10, Section 5) in

a) the space of functions on the set of vertices of the cube;

b) the space of functions on the set of edges of the tetrahedron.

5.* Let $T: A_5 \to \mathrm{GL}(V)$ be a three-dimensional irreducible representation. Decompose $\mathrm{L}(V)$ into a direct sum of minimal invariant subspaces with respect to the representation TT'.

6. Let s be an arbitrary action of the finite group G on the set X. Let $x_1, \dots, x_m$ be representatives of the distinct orbits and $H_1, \dots, H_m$ the corresponding isotropy subgroups. Prove that

$$s_* \simeq L^{H_1} + \dots + L^{H_m}.$$

7. Let T be an n-dimensional irreducible complex representation of the finite group G. Prove that

$$\sum_{g \in G} \overline{\chi_T(g)} T(g) = \frac{|G|}{n} \varepsilon.$$

(*Hint.* Use Schur's Lemma to prove that the operator appearing on the left-hand side is scalar, and then compute its trace.)

8.* Prove that the dimension of any irreducible complex representation of the finite group G divides the order of G. (*Hint.* Use Exercise 7 to show that the trace of any power of the operator $\frac{|G|}{n}\varepsilon$ is an integer.)

III. Representations of Compact Groups

If we leave the framework of pure algebra and decide to consider continuous homomorphisms, continuous representations, and so on, we discover that compact topological groups are in many respects similar to finite groups. In this chapter we examine the simplest examples of connected compact groups and their linear representations, and we generalize to compact linear groups the main theorems on matrix elements proven in the preceding chapter for finite groups. In the next chapter we will discuss the basic method of the theory of Lie groups and then use it, in particular, to describe all linear representations of the simplest compact groups. All this, however, can serve only as an introduction to the present day theory of linear representations of compact groups, the most important parts of which are

1) the classification of linear representations of compact connected Lie groups, and

2) the Peter-Weyl Theorem on the completeness of the set of matrix elements of an arbitrary compact group.

For the classification of representations of compact Lie groups, we send the reader to the books [1] and [11], among others. Expositions of the proof of the Peter-Weyl Theorem are given in [7] and [10].

7. The groups SU_2 and SO_3

The simplest connected compact group is the commutative group $\mathbf{T}$, the unit circle. Among the noncommutative connected compact groups, the simplest are the intimately related groups SU_2 and SO_3, to which this section is devoted.

7.1. SU_2. It is readily seen that

$$SU_2 = \left\{ \begin{pmatrix} z_1 & z_2 \\ -\bar{z}_2 & \bar{z}_1 \end{pmatrix} \mid z_1, z_2 \in \mathbf{C},\ |z_1|^2 + |z_2|^2 = 1 \right\}.$$

E.B. Vinberg, *Linear Representations of Groups*, Modern Birkhäuser Classics,
DOI 10.1007/978-3-0348-0063-1_4,

Consider the set of matrices

$$\mathbf{H} = \left\{ \begin{pmatrix} z_1 & z_2 \\ -\bar{z}_2 & \bar{z}_1 \end{pmatrix} \mid z_1, z_2 \in \mathbf{C} \right\}.$$

It is a four-dimensional real subspace of the vector space $\mathrm{L}_2(\mathbf{C})$, and enjoys the following remarkable properties:

1) $\mathbf{H}$ is closed under multiplication, i.e., it is a real subalgebra of the algebra $\mathrm{L}_2(\mathbf{C})$;

2) the determinant is a positive definite quadratic function on $\mathbf{H}$:

$$\det \begin{pmatrix} z_1 & z_2 \\ -\bar{z}_2 & \bar{z}_1 \end{pmatrix} = |z_1|^2 + |z_2|^2 = x_0^2 + x_1^2 + x_2^2 + x_3^2,$$

where we put $z_1 = x_0 + ix_1$, $z_2 = x_2 + ix_3$ (with $x_0, x_1, x_2, x_3 \in \mathbf{R}$);

3) $\mathbf{H}$ contains, together with any nonnull matrix, its inverse:

$$\begin{pmatrix} z_1 & z_2 \\ -\bar{z}_2 & \bar{z}_1 \end{pmatrix}^{-1} = \frac{1}{|z_1|^2 + |z_2|^2} \begin{pmatrix} \bar{z}_1 & -z_2 \\ \bar{z}_2 & z_1 \end{pmatrix} \in \mathbf{H}.$$

Properties 1) and 3) show that $\mathbf{H}$ is a division algebra. It is called (independently of its specific matrix realization) the QUATERNION ALGEBRA.

As a basis for the quaternion algebra one can take the elements e_0, e_1, e_2, and e_3, which in our model are represented by the matrices

$$\begin{pmatrix} 1 & 0 \\ 0 & 1 \end{pmatrix}, \quad \begin{pmatrix} i & 0 \\ 0 & -i \end{pmatrix}, \quad \begin{pmatrix} 0 & 1 \\ -1 & 0 \end{pmatrix}, \quad \text{and} \quad \begin{pmatrix} 0 & i \\ i & 0 \end{pmatrix}.$$

They satisfy the following multiplication rules:

$$\begin{aligned} &e_0 e_i = e_i e_0 = e_i, \quad \text{for } i = 0, 1, 2, 3; \\ &e_i^2 = -e_0, \quad \text{for } i = 1, 2, 3; \\ &e_i e_j = -e_j e_i = e_k, \quad \text{if } (i, j, k) \text{ is an even permutation of } (1, 2, 3). \end{aligned}$$

Taking the determinant $\det A$ of a matrix A as its squared magnitude (i.e., the inner product of A with itself), we turn $\mathbf{H}$ into a *four-dimensional Euclidean space.* The group SU_2 consists of the matrices $A \in \mathbf{H}$ satisfying $(A, A) = 1$, i.e., it is the *unit sphere* in $\mathbf{H}$. This implies, in particular, that SU_2 is connected.

Since the determinant of a product of matrices is the product of the determinants of the factors, right or left multiplication by any matrix $A \in \mathrm{SU}_2$ is an *orthogonal transformation* of the space $\mathbf{H}$:

$$(AX, AX) = (XA, XA) = (A, A)(X, X) = (X, X).$$

It follows that the *invariant integration* (see 5.2) on SU_2 can be defined as the usual integration on the three-dimensional sphere with a suitable normalizing factor.

7.2. The homomorphism $SU_2 \to SO_3$**.** To establish the relation between the groups SU_2 and SO_3 we consider the *three-dimensional Euclidean space* $\mathbf{E}$ of the 2×2 traceless Hermitian matrices, i.e., matrices of the form

$$X = \begin{pmatrix} x_1 & x_2 + ix_3 \\ x_2 - ix_3 & -x_1 \end{pmatrix} \qquad (x_1, x_2, x_3 \in \mathbf{R}),$$

with the inner product

$$(X, X) = -\det X = x_1^2 + x_2^2 + x_3^2.$$

Notice that $i\mathbf{E}$ coincides with the orthogonal complement to the unit element e_0 in the algebra $\mathbf{H}$.

Next, we consider the linear representation P of SU_2 in $\mathbf{E}$ defined by the formula

$$P(A)X = AXA^{-1} \qquad (A \in SU_2,\ X \in \mathbf{E}).$$

If one regards X as the matrix of a Hermitian operator in an orthonormal basis, then AXA^{-1} is the matrix of the same operator in another such basis. Consequently, AXA^{-1} is again a traceless Hermitian matrix, i.e., an element of $\mathbf{E}$ (which of course can also be verified by a direct calculation).

It follows from the equality $\det(AXA^{-1}) = \det X$ that the transformations $P(A)$ are orthogonal. This implies a priori that $\det P(A)$ can assume only the values ± 1. However, the connectedness of the group SU_2 and the continuity of the function $A \to \det P(A)$ on it force the equality $\det P(A) = 1$ for all $A \in SU_2$. We have thus obtained a homomorphism

$$P\colon SU_2 \to SO_3. \tag{1}$$

Theorem. *The homomorphism (1) is surjective. Its kernel consists of the two matrices E and $-E$.*

For the proof we need two lemmas.

Lemma 1. *Let F be a subgroup of* SO_3 *with the following properties:*

1) F acts transitively on the unit sphere, i.e., any given unit vector can be taken into any other such vector by a suitable transformation belonging to F;

2) there is an axis such that F contains all rotations around that axis.

Then $F = SO_3$.

Proof. Suppose that F contains all rotations around the axis $\langle e \rangle$, where $|e| = 1$. Let g be an arbitrary orthogonal transformation with determinant 1. By assumption, there is an $f \in F$ such that $fe = ge$ or, equivalently, $f^{-1}ge = e$. The transformation $f^{-1}g$ is therefore a rotation around $\langle e \rangle$ and so, by 2), $f^{-1}g \in F$. Consequently, $g = f(f^{-1}g) \in F$ also. □

Lemma 2. *For every* $X \in \mathbf{E}$ *there is an* $A \in \mathrm{SU}_2$ *such that* $AXA^{-1} = cH$, *where* $H = \begin{pmatrix} 1 & 0 \\ 0 & -1 \end{pmatrix}$ *and* $c > 0$.

PROOF. Think of X as the matrix of a Hermitian operator in an orthonormal basis. It is known that for every Hermitian operator X there exists an orthonormal basis consisting of eigenvectors of X. Hence, there exists a unitary matrix B such that

$$BXB^{-1} = cH \qquad (c \in \mathbf{R}).$$

Permuting the basis vectors, if necessary, one can arrange that $c > 0$. Now set

$$A = (\det B)^{-1/2} B.$$

Since $|\det B| = 1$, A is a unitary matrix, and so $A \in \mathrm{SU}_2$. Moreover, it is clear that

$$AXA^{-1} = BXB^{-1} = cH.$$

□

PROOF OF THE THEOREM. To establish the surjectivity of P we apply Lemma 1 to the subgroup $F = P(\mathrm{SU}_2)$. That F possesses property 1) follows from Lemma 2. To verify property 2), we calculate $P(A(z))$, where

$$A(z) = \begin{pmatrix} z & 0 \\ 0 & z^{-1} \end{pmatrix} \qquad (z \in \mathbf{C}, \quad |z| = 1). \tag{2}$$

We have

$$P(A(z)) \begin{pmatrix} x_1 & x_2 + ix_3 \\ x_2 - ix_3 & -x_1 \end{pmatrix} = \begin{pmatrix} x_1 & z^2(x_2 + ix_3) \\ z^{-2}(x_2 - ix_3) & -x_1 \end{pmatrix}, \tag{3}$$

which shows that for $z = \mathrm{e}^{it}$ the operator $P(A(z))$ is the rotation through an angle of $2t$ around the axis $\langle H \rangle$.

It remains to find the kernel of P. If $P(A) = \varepsilon$, then, in particular, $P(A)H = H$, which implies that A is a diagonal matrix, i.e., of the form $A(z)$. But, as formula (3) shows, $P(A(z)) = \varepsilon$ if and only if $z = \pm 1$. Hence, $A = \pm E$, as claimed. □

7.3. SO_3. It follows from the theorem just proved that

$$\mathrm{SO}_3 \simeq \mathrm{SU}_2 / \{E, -E\}. \tag{4}$$

As a consequence, every linear representation of SO_3 is obtained by factoring a linear representation of SU_2 (see 3.8). However, in order to be able to make the same assertion about continuous linear representations, we have to give a topological meaning to the isomorphism (4).

Generally, let G be a topological group, N a closed normal subgroup of G, and $\pi: G \to G/N$ the canonical homomorphism. The quotient group G/N is endowed with a topology in the following standard manner: a subset $B \subset G/N$ is declared closed if and only if its preimage $\pi^{-1}(B)$ is closed in G. Under this convention a function f on G/N is continuous if and only if the function $\pi^* f$ on G given by

$$(\pi^* f)(x) = f(\pi x) \qquad (x \in G)$$

is continuous. Consequently, a linear representation T of the quotient group G/N is continuous if and only if the representation $T \circ \pi$ of G is. That is to say, the continuous representations of G/N are obtained by factoring the continuous representations of G whose kernels contain N.

Lemma. *Let p be a continuous homomorphism of the compact topological group G onto the topological group F. Then, as a topological group, F is isomorphic to the quotient group G/N, where N is the kernel of p. More precisely, the map $\bar{p}: G/N \to F$ sending each coset gN into the element $p(g)$ is an isomorphism of topological groups.*

Proof. The usual homomorphism theorem asserts that $\bar{p}$ is a group isomorphism. If π denotes the canonical homomorphism $G \to G/N$, we have the commutative diagram

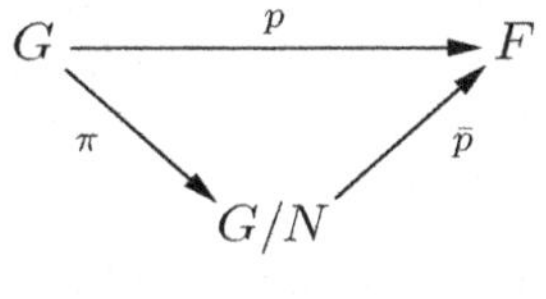

Let us check that $\bar{p}$ is continuous. Let B be a subset of F. Then $\pi^{-1}(\bar{p}^{-1}(B)) = p^{-1}(B)$. If B is closed in F, then, thanks to the continuity of the map p, $p^{-1}(B)$ is closed in G. By the definition of the topology on G/N, the latter means precisely that $\bar{p}^{-1}(B)$ is closed in G/N.

It remains to check that $\bar{p}^{-1}$ is continuous, or equivalently, that $\bar{p}$ is a closed map. Let A be a closed subset of G/N. We have $\bar{p}(A) = p(\pi^{-1}(A))$. Since the set $\pi^{-1}(A)$ is closed in G, it is compact. Its image under p is also compact and hence closed in F, as we needed to show. $\square$

Applying the lemma to the homomorphism P constructed in 7.2, we conclude that (4) is an isomorphism of topological groups. In particular, as a topological space, SO_3 is the three-dimensional sphere with each pair of diametrically opposed points identified, i.e., the *three-dimensional real projective space*. We also obtain the following description of the continuous linear representations of SO_3.

Proposition. *The continuous representations of the group* SO_3 *are obtained by factoring those continuous representations of the group* SU_2 *whose kernels contain* $-E$.

7.4. We next construct a series of *irreducible complex representations* of SU_2 and SO_3.

Let V_n denote the space of forms (homogeneous polynomials) of degree n in the variables u_1 and u_2. We define a representation Φ of the group $SL_2(\mathbf{C})$ in V_n by the rule

$$(\Phi_n(A)f)(u) = f(uA), \tag{5}$$

where $u = (u_1, u_2) \in \mathbf{C}^2$, $A \in SL_2(\mathbf{C})$, and $f \in V_n$. In more detail, if $A = [a_{ij}]$, then

$$(\Phi_n(A)f)(u_1, u_2) = f(a_{11}u_1 + a_{21}u_2, a_{12}u_1 + a_{22}u_2). \tag{6}$$

This definition agrees with the general rule by which transformations act on functions if one regards u_1 and u_2 as coordinates on the space U dual to the space of column vectors in which $SL_2(\mathbf{C})$ acts naturally, and one accordingly regards V_n as a space of functions on U. In this interpretation, the space of column vectors itself and the identity representation of $SL_2(\mathbf{C})$ are canonically isomorphic to V_1 and the representation Φ_1 respectively.

Notice that Φ_0 is the trivial one-dimensional representation of $SL_2(\mathbf{C})$.

The restriction of the representation Φ_n to the subgroup $SU_2 \subset SL_2(\mathbf{C})$ will be denoted by the same symbol Φ_n.

Proposition. *The representation* Φ_n *of* SU_2 *is irreducible.*

Proof. Let $\mathbf{T}$ be the subgroup of SU_2 consisting of the diagonal matrices $A(z)$ (see 7.2). We first determine which of the subspaces of V_n are invariant under $\mathbf{T}$. It is a straightforward matter to check that

$$\Phi_n(A(z))u_1^{n-k}u_2^k = z^{n-2k}u_1^{n-k}u_2^k. \tag{7}$$

It follows that V_n splits into the direct sum of the one-dimensional $\mathbf{T}$-invariant subspaces $\langle u_1^{n-k}u_2^k\rangle$, $k = 0, 1, \ldots, n$. Moreover, the corresponding subrepresentations of $\mathbf{T}$ are pairwise nonisomorphic. By Theorem 2 of 4.1, every $\mathbf{T}$-invariant subspace of V_n is a linear span of a number of monomials $u_1^{n-k}u_2^k$.

Now let W be an arbitrary nonnull SU_2-invariant subspace of V_n. By the foregoing discussion, W contains a monomial $u_1^{n-k}u_2^k$. Pick any nondiagonal matrix $A_0 \in SU_2$ and let it act on $u_1^{n-k}u_2^k$. It is readily seen that the coefficient of u_1^n in the form $f_0 = \Phi_n(A_0)u_1^{n-k}u_2^k$ is different from zero. Since $f_0 \in W$ and W is spanned by monomials, it follows that $u_1^n \in W$.

Analogously, considering the form $\Phi_n(A_0)u_1^n$, we remark that all its coefficients are different from zero. We thus conclude that all monomials belong to W, i.e., $W = V_n$, which completes the proof. □

Obviously

$$\Phi_n(-E) = (-1)^n\varepsilon. \tag{8}$$

Hence, $-E$ belongs to the kernel of Φ_n if and only if n is even. For such values of n the representation Φ_n of SU_2 can be factored with respect to the normal subgroup $\{E, -E\}$, thereby yielding an irreducible representation of SO_3 that we will denote by Ψ_n.

Thus, for each integer $n \geq 0$ we have constructed an irreducible $(n+1)$-dimensional representation Φ_n of SU_2, and for each even $n \geq 0$, an irreducible $(n+1)$-dimensional representation Ψ_n of SO_3. In Section 11 we will show that *this is a complete list of the continuous irreducible complex representations of the groups* SU_2 *and* SO_3. (See also the Exercises in Section 8.)

Questions and Exercises

1. For arbitrary $A, B \in SU_2$ put

$$R(A,B)X = AXB^{-1} \qquad (X \in \mathbf{H}).$$

Show that R is a homomorphism of $SU_2 \times SU_2$ onto SO_4, and find its kernel.

2.* Let P be the linear representation of SU_2 constructed in 7.2. Construct an explicit isomorphism of the representations $P_{\mathbf{C}}$ and Φ_2 of SU_2.

3. Prove that any central function f on SU_2 is uniquely determined by its restriction to the subgroup

$$\mathbf{T} = \left\{ A(z) = \begin{pmatrix} z & 0 \\ 0 & z^{-1} \end{pmatrix} \mid z \in \mathbf{C},\ |z| = 1 \right\},$$

and that $f(A(z)) = f(A(z^{-1}))$.

4. Compute the restriction to $\mathbf{T}$ of the character χ_n of the representation Φ_n of SU_2.

5. Prove that the linear span of the functions

$$\phi_n(z) = \chi_n(A(z)) \qquad (z \in \mathbf{C},\ |z| = 1)$$

coincides with the space of all functions ϕ on the unit circle which can be written as polynomials in z and $\bar{z}$, and which satisfy the condition $\phi(\bar{z}) = \phi(z)$.

6. Let f be a continuous central function on SU_2. Show that

$$\int_{\mathrm{SU}_2} f(x)\,dx = \frac{2}{\pi}\int_0^{\pi} f(A(\mathrm{e}^{it}))\,\sin^2 t\,dt.$$

8. Matrix Elements of Compact Groups

In this section we generalize the main theorems established in Chapter II for finite groups to compact linear groups.

We shall consider only (continuous) *complex* linear representations. Recall that every complex representation of a compact group is unitary, and hence completely reducible (see Section 2).

8.1. Let X be a compact topological space on which integration is defined, i.e., there is given a positive linear functional

$$f \mapsto \int_X f(x)\,dx$$

on the space of continuous real-valued functions on X. We extend the integral by linearity to continuous complex-valued functions. Specifically, if $f = g+ih$, where g, h are continuous real-valued functions, we put

$$\int_X f(x)\,dx = \int_X g(x)\,dx + i\int_X h(x)\,dx.$$

Now, in the space of continuous complex functions on X we define a Hermitian inner product by the rule

$$(f_1, f_2) = \int_X f_1(x)\overline{f_2(x)}\,dx.$$

We let $C_2(X)$ denote the resulting (generally speaking, infinite-dimensional) Hermitian space.

A countable or finite orthogonal set of functions $f_1, f_2, \ldots \in C_2(X)$ is said to be COMPLETE if its linear span is dense in $C_2(X)$ in the topology defined by the Hermitian metric. If this is the case, then every function $f \in C_2(X)$ can be expressed as the sum (in the topology of $C_2(X)$) of a Fourier series:

$$f = \sum_k a_k f_k, \qquad \text{where } a_k = \frac{(f, f_k)}{(f_k, f_k)}.$$

For a compact topological group G we define the space $C_2(G)$ using the normalized invariant integration on G (see 2.5).

In particular, if G is a finite group, then $C_2(G) = \mathbf{C}[G]$, and the inner product in $C_2(G)$ is identical with that defined in Section 6.

In any case, *the inner product of $C_2(G)$ is invariant under left and right translations,* i.e., under the two-sided regular representation Reg of the group $G \times G$ in $C_2(G)$. In fact,

$$\begin{aligned}(\mathrm{Reg}(g_1, g_2)f_1, \mathrm{Reg}(g_1, g_2)f_2) &= \int_G f_1(g_2^{-1}xg_1)\overline{f_2(g_2^{-1}xg_1)}\,dx \\ &= \int_G f_1(x)\overline{f_2(x)}\,dx = (f_1, f_2).\end{aligned}$$

8.2. The simplest example of an infinite compact group is $\mathbf{T} = \{z \in \mathbf{C}^* \mid |z| = 1\}$. It possesses the infinite series of one-dimensional irreducible representations

$$z \mapsto z^n \qquad (n \in \mathbf{Z}).$$

It is known that the functions z^n ($n = 0, \pm1, \pm2, \ldots$) constitute a complete orthogonal set in the space $C_2(\mathbf{T})$. This fact lies at the foundation of the theory of classical Fourier series.

The reader should compare this with Theorem 1 of Section 6, according to which the matrix elements of the irreducible representations of a finite group G form a complete orthogonal set in $C_2(G)$.

The facts stated above also hold in a far more general context. Specifically, we have the following result.

Theorem 1. *Let G be a compact topological group. Then the matrix elements of the irreducible representations of G form a complete orthogonal set in the space $C_2(G)$. Moreover, the inner product of any such matrix element with* *itself is equal to* n^{-1}, *where* n *is the dimension of the corresponding irreducible representation.*

[As in Section 6, let us agree that for each irreducible representation the matrix elements are defined relative to a basis which is orthonormal with respect to an invariant inner product in the representation space.]

The theorem contains two independent assertions: one about the orthogonality and the second about the completeness of the set of matrix elements.

The *proof of the orthogonality* and the calculation of the norms of the matrix elements is carried out in exactly the same manner as for finite groups (see the proof of Theorem 1, Section 6), with integration taking the place of summation over the group.

The *proof of the completeness* of the set of matrix elements is given below only for linear groups.

8.3. Let G be a compact subgroup of $\mathrm{GL}_n(\mathbf{C})$. Let a_{ij} denote the matrix elements of the identity representation of G, and let $\mathbf{C}[G]$ stand for the space of all functions expressible as polynomials in a_{ij} and $\bar{a}_{ij}$. By Weierstrass's Theorem, any continuous function on G can be uniformly approximated by functions in $\mathbf{C}[G]$. Consequently, $\mathbf{C}[G]$ *is a dense subspace of* $C_2(G)$. If G is finite, then $\mathbf{C}[G]$ coincides with the space of all functions on G, and so in this case our notation agrees with that used in Chapter II.

Further, let $\{T_k\}$ be the (generally speaking, infinite) collection of all irreducible representations of the group G, and let $\mathrm{M}(T_k)$ denote the space of matrix elements of T_k (see 4.5).

Theorem 2. *For every compact linear group*

$$\mathbf{C}[G] = \sum \mathrm{M}(T_k),$$

where the sum is direct. (Cf. Theorem 2, 5.2.)

Proof. We let $\mathbf{C}[G]_m$ denote the space of all functions on G that can be written as polynomials of degree at most m in the variables a_{ij} and $\bar{a}_{ij}$. It is obviously finite dimensional. Since under (left and right) translations on G the matrix elements a_{ij} transform linearly, the function space $\mathbf{C}[G]_m$ is both left and right invariant.

We now repeat the proof of Theorem 2 of 5.2, with the difference that instead of the representation R, which may be infinite-dimensional, we take its finite-dimensional subrepresentation R_m in the invariant subspace $\mathbf{C}[G]_m$. We then have the inclusions

$$\mathbf{C}[G]_m \subset \mathrm{M}(R_m) \subset \sum \mathrm{M}(T_k).$$

It is clear that $\mathbf{C}[G]_m$ is actually contained in the sum of a finite number of subspaces $\mathrm{M}(T_k)$. Since $\mathbf{C}[G]_m$ is invariant under the representation Reg, and since $\mathrm{Reg}_{\mathrm{M}(T_k)} \not\simeq \mathrm{Reg}_{\mathrm{M}(T_\ell)}$ if $k \neq \ell$ (Corollary 4 to Theorem 7, 4.6), it follows that $\mathbf{C}[G]_m$ is simply the sum of a finite number of subspaces $\mathrm{M}(T_k)$ (Theorem 2, 4.1).

The space $\mathbf{C}[G]$ is the union of all $\mathbf{C}[G]_m$. Consequently it, too, is a sum of (generally speaking, infinitely many) subspaces $\mathrm{M}(T_k)$. Suppose that $\mathrm{M}(T_s) \not\subset \mathbf{C}[G]$ for some s. Then the subspace $\mathrm{M}(T_s)$ is orthogonal to $\mathbf{C}[G]$, and hence to all continuous functions on G, contradicting the fact that $\mathrm{M}(T_s)$ itself consists of continuous functions. This completes the proof of the theorem. □

Corollary 1. *The matrix elements of the irreducible linear representations of G form an orthogonal basis of* $\mathbf{C}[G]$. □

Corollary 2. *The matrix elements of the irreducible linear representations of G form a complete orthogonal set in $C_2(G)$.* □

We have thus proved Theorem 1 for compact linear groups.

Theorems 1 and 2 permit us to develop the theory of characters of compact groups, the basic assertions of which are the same as in the theory of characters of finite groups (Corollaries 1–3 to Theorem 1, Section 6). We shall not pursue this further, however.

8.4. For an arbitrary subgroup H of the compact linear group G, the space

$$\mathbf{C}[G]^H = \{\, f \in \mathbf{C}[G] \mid f(gh) = f(g) \ \text{ for all } g \in G,\, h \in H \,\}$$

is described exactly as in Theorem 4 of 5.6 as

$$\mathbf{C}[G]^H = \sum \mu_k (V_k^H \otimes V_k'), \tag{1}$$

where V_k is the representation space of T_k. It follows that

$$L^H \simeq \sum m_k T_k, \quad \text{where } m_k = \dim V_k^H. \tag{2}$$

To pass to arbitrary continuous functions we use the following lemma.

Lemma. *The closure of* $\mathbf{C}[G]^H$ *in* $C_2(G)$ *coincides with* $C_2(G)^H$.

PROOF. Every function $f \in C_2(G)$ can be uniquely written as the sum (understood in the topology of $C_2(G)$) of a series

$$f = \sum f_k, \qquad f_k \in \mathrm{M}(T_k).$$

Accordingly,

$$R(h)f = \sum R(h)f_k, \qquad R(h)f_k \in \mathrm{M}(T_k).$$

This implies that

$$f \in C_2(G)^H \iff (f_k \in \mathrm{M}(T_k)^H \text{ for all } k),$$

and so

$$C_2(G)^H = \overline{\sum \mathrm{M}(T_k)^H} = \overline{\mathbf{C}[G]^H}.$$

□

The meaning of the lemma is that the orthogonal basis of the space $\mathbf{C}[G]^H$ provided by the matrix elements of the representations T_k is a complete orthogonal set in $C_2(G)^H$.

Questions and Exercises

1. How does the proof of the orthogonality of the functions z^n on the unit circle go if one proceeds as in the proof of Theorem 9, 4.7?

2. Prove that every irreducible continuous representation of $\mathbf{T}$ has the form $z \mapsto z^n$.

3. Find all irreducible continuous representations of the k-dimensional torus $\mathbf{T}^k$.

4. Prove that a compact linear group has at most countably many nonisomorphic irreducible representations.

5. Prove that the characters of the irreducible representations of any compact linear group G form a complete orthonormal set in the space of continuous central functions on G.

6. Let G be a compact linear group and H a subgroup of G. Construct from matrix elements of irreducible representations of G an orthogonal basis of the space $\mathbf{C}[G]^H$.

7. Using Exercises 3–5 of Section 7, show that the characters of the representations Φ_n of SU_2 constructed in 7.4 constitute a complete orthogonal set in the space of continuous central functions on SU_2. Conclude from this that every irreducible representation of SU_2 is isomorphic to one of the representations Φ_n.

8. Use the theory of characters to derive the following relation for the representations of SU_2: if $m \geq n$, then

$$\Phi_m \Phi_n \simeq \Phi_{m-n} + \Phi_{m-n+2} + \ldots + \Phi_{m+n}.$$

9. Using Exercise 7, show that every irreducible representation of SO_3 is isomorphic to one of the representations Ψ_n constructed in 7.4.

10. Let R and S be representations of SU_2 such that $R|_{\mathbf{T}} \simeq S|_{\mathbf{T}}$. Show that $R \simeq S$. (See Exercise 3, Section 7.)

9. The Laplace Spherical Functions

9.1. We consider the following general problem. Suppose we are given a transitive action $s: G \to S(X)$ of the compact topological group G on the topological space X. We shall assume that s is continuous, i.e., the map

$$G \times X \to X, \qquad (g, x) \mapsto s(g)x,$$

is continuous. Under these circumstances, can we decompose the space of continuous functions on X into a topological direct sum of finite-dimensional G-invariant subspaces? If yes, how can the linear representations of G in these subspaces be described?

Let H be the isotropy subgroup of some point $o \in X$. It is a closed subgroup of G. Proceeding as in 5.5, we construct the bijective map

$$\bar{p}: G/H \to X, \qquad gH \mapsto go,$$

which commutes with the actions of the group G, where one assumes, as usual, that G acts on G/H by left translations.

The left coset space G/H is endowed with a topology as follows: a subset $B \subset G/H$ is said to be closed if and only if its preimage under the canonical projection $\pi: G \to G/H$ is closed.

Lemma. *$\bar{p}$ is a homeomorphism.*

The proof is analogous to that of the Lemma of 7.3. □

Thus, the space X can be identified with G/H. Under this identification s becomes the action l^H, and the continuous functions on X become the continuous functions on G/H or, equivalently, the continuous functions on G which are constant on the left cosets of H in G. Accordingly, the representation s_* of G in the space of continuous functions on X is isomorphic to the representation L^H of G in the space $C_2(G)^H$. The decomposition of L^H obtained in 8.4 permits us to answer the questions formulated above, provided that the irreducible representations of the group G are known.

Specifically, let $(e_1, \dots, e_n)$ be an orthonormal basis of the space V of the irreducible representation T of G, chosen so that $(e_1, \dots, e_m)$ is a basis of the subspace V^H. Let T_{ij} be the matrix elements of T in this basis. Then for each $j = 1, \dots, m$ the functions $T_{1j}, \dots, T_{nj}$ are constant on the left cosets of H in G. Regarded as functions on X, they form a basis for a minimal G-invariant subspace of functions in which a representation isomorphic to T' is realized. The functions constructed in this manner for all irreducible representations T of G form a complete orthogonal set in the space of continuous functions on X. They will be referred to as SPHERICAL FUNCTIONS to emphasize the analogy with Laplace's spherical functions, to which the remaining part of this section is devoted.

9.2. A particular case of the problem considered in the preceding subsection is that of *decomposing the space of continuous functions on the two-dimensional sphere*

$$S = \{ (x_1, x_2, x_3) \in \mathbf{R}^3 \mid x_1^2 + x_2^2 + x_3^2 = 1 \}$$

into a topological direct sum of finite-dimensional subspaces invariant under the group SO_3 *of rotations of* $\mathbf{R}^3$. We solve this problem here without resorting to the foregoing general considerations.

Let $C_2(S)$ denote the Hermitian space of all continuous complex functions on S with the inner product defined as in 8.1. As the (usual) integration on S is invariant under SO_3, the inner product in $C_2(S)$ is also invariant under SO_3.

Fix the point $o = (0, 0, 1)$, the "north pole" on S. The isotropy subgroup of o, that is, the group of rotations around the x_3-axis, is isomorphic to SO_2, and by a slight abuse of notation we shall denote it by SO_2.

Lemma 1. *In every nonnull finite-dimensional* SO_3*-invariant subspace* $U \subset C_2(S)$ *there is a nonnull* SO_2*-invariant function.*

PROOF. First of all, let us show that U contains functions that do not vanish at the point o. Let $f \in U$ be an arbitrary nonnull function, and let $x \in S$ be such that $f(x) \neq 0$. There is a rotation $g \in \mathrm{SO}_3$ such that $gx = o$, and then $(g_* f)(o) = f(x) \neq 0$.

Now consider the subspace

$$U_0 = \{ f \in U \mid f(o) = 0 \}.$$

By the foregoing, it has codimension one. It is clearly SO_2-invariant. Hence, its orthogonal complement $U_0^\perp$ is also SO_2-invariant.

Now let $f_0 \in U_0^\perp$ be an arbitrary nonnull function. Rotations around the x_3-axis can affect f_0 only by a multiplicative constant. However, since the point o is fixed, that constant is in fact equal to 1. Hence, f_0 is the desired function. □

9.3. Let A denote the algebra of polynomials with complex coefficients in the coordinates x_1, x_2, x_3. Its elements will be regarded as functions on $\mathbf{R}^3$. It is plain that the algebra A is SO_3-invariant. Moreover, it splits into a direct sum of finite-dimensional invariant subspaces:

$$A = A_0 \oplus A_1 \oplus A_2 \oplus \ldots,$$

where A_m designates the space of homogeneous polynomials of degree m. The monomials $x_1^{k_1} x_2^{k_2} x_3^{k_3}$ provide a basis for A. We endow A with a Hermitian inner product chosen so that this basis is orthogonal and

$$(x_1^{k_1} x_2^{k_2} x_3^{k_3}, x_1^{k_1} x_2^{k_2} x_3^{k_3}) = k_1! k_2! k_3!\,.$$

Lemma 2. *The operator of multiplication by x_i in A is the adjoint of the differentiation operator $\partial/\partial x_i$ with respect to the inner product $(\,\cdot\,,\,\cdot\,)$.*

PROOF. For the sake of definiteness, let $i = 1$. We have to show that $(\partial u/\partial x_1, v) = (u, x_1 v)$ for any two polynomials u, v. It suffices to check this for monomials u and v. Let $u = x_1^{k_1} x_2^{k_2} x_3^{k_3}$. If $v \neq x_1^{k_1-1} x_2^{k_2} x_3^{k_3}$, then both sides of the needed equality vanish. If now $v = x_1^{k_1-1} x_2^{k_2} x_3^{k_3}$, then we get $k_1! k_2! k_3!$ in both sides. □

Corollary. *The operator of multiplication by $r^2 = x_1^2 + x_2^2 + x_3^2$ is the adjoint to the Laplace operator*

$$\Delta = \frac{\partial^2}{\partial x_1^2} + \frac{\partial^2}{\partial x_2^2} + \frac{\partial^2}{\partial x_3^2}.$$

□

Notice that $r^2A_m \subset A_{m+2}$ and $\Delta A_m \subset A_{m-2}$.

The functions annihilated by the operator Δ are called HARMONIC. We let H denote the space of harmonic polynomials. Obviously $H = \sum_m H_m$, where $H_m = H \cap A_m$. Since the kernel of any linear operator coincides with the orthogonal complement to the range of its adjoint,

$$(1) \qquad A_m = H_m \oplus r^2 A_{m-2}, \qquad m = 0, 1, 2, \ldots.$$

This yields the following direct sum decomposition of the space A_m:

$$(2) \qquad A_m = H_m \oplus r^2 H_{m-2} \oplus r^4 H_{m-4} \oplus \cdots.$$

Since both the Laplace operator and the operator of multiplication by r^2 commute with rotations, the summands in this decomposition are SO_3-invariant. In point of fact, they are *minimal invariant subspaces,* as we next show.

9.4. Let ρ denote the operation of *restriction* of functions from $\mathbf{R}^3$ to the sphere S. Set

$$\mathbf{C}[S] = \rho(A).$$

It is plain that the map

$$\rho\colon A \to \mathbf{C}[S]$$

is linear and commutes with the actions of SO_3 on A and $\mathbf{C}[S]$.

It follows from Weierstrass's Theorem that $\mathbf{C}[S]$ is a dense subspace of $C_2(S)$ (cf. 8.3).

Lemma 3. $\operatorname{Ker} \rho \cap A_m = 0$.

PROOF. In fact, any homogeneous polynomial that vanishes identically on S vanishes identically in $\mathbf{R}^3$, too. □

Theorem 1. *(2) is a decomposition of A_m into a direct sum of minimal* SO_3-*invariant subspaces.*

PROOF. By Lemma 3, ρ maps A_m isomorphically onto $\rho(A_m)$. By Lemma 1, the number of components in a direct sum decomposition of A_m into minimal invariant subspaces does not exceed $\dim A_m^{\mathrm{SO}_2}$. We next show that $\dim A_m^{\mathrm{SO}_2} = [m/2] + 1$ (Lemma 4 below; here $[\;]$ stands for integer part). Since the number of components in the decomposition (2) is also equal to $[m/2] + 1$, we conclude that the latter are minimal invariant subspaces. □

Let us parametrize SO_2 by complex numbers of modulus one. Namely, for each $z = e^{it}$, $t \in \mathbf{R}$, we let $h(z)$ denote the rotation through angle t around the x_3-axis.

Lemma 4. *The space A_m admits a basis of joint eigenfunctions of the transformations $h(z) \in SO_2$. The corresponding eigenvalues have the form z^k, for $k = 0, \pm 1, \pm 2, \ldots, \pm m$. The multiplicity of the eigenvalue z^k is $[\frac{1}{2}(m - |k|)] + 1$. In particular,*

$$\dim A_m^{SO_2} = \left[\frac{m}{2}\right] + 1.$$

PROOF. Put

$$u = x_1 - ix_2, \qquad \bar{u} = x_1 + ix_2.$$

It is readily seen that

$$h(z)_* u = zu, \qquad h(z)_* \bar{u} = z^{-1}\bar{u}.$$

Any polynomial in A_m is uniquely expressible as a linear combination of monomials $u^p \bar{u}^q x_3^\ell$, where $p + q + \ell = m$. From the above formulas it follows that

$$h(z)_* u^p \bar{u}^q x_3^\ell = z^{p-q} u^p \bar{u}^q x_3^\ell,$$

i.e., the monomials $u^p \bar{u}^q x_3^\ell$ are joint eigenfunctions for all transformations in SO_2 with corresponding eigenvalues z^{p-q}. Since $|p - q| \le p + q = m - \ell$, we have that $|p - q| \le m$. For $p - q = k$ the exponent ℓ can assume the values $m - |k|, m - |k| - 2, m - |k| - 4, \ldots$. Hence, the number of monomials $u^p \bar{u}^q x_3^\ell$ with $p - q = k$ is $[\frac{1}{2}(m - |k|)] + 1$. □

9.5. We can now formulate the main result of this section.

Theorem 2. *The space $\mathbf{C}[S]$ decomposes into the orthogonal direct sum of the minimal SO_3-invariant subspaces $U_m = \rho(H_m)$, $m = 0, 1, 2, \ldots$. The subspace U_m has dimension $2m + 1$. It admits an orthogonal basis $(Y_{m,0}, Y_{m,\pm 1}, \ldots, Y_{m,\pm m})$ consisting of joint eigenfunctions of the transformations $h(z) \in SO_2$. The eigenvalue corresponding to $Y_{m,k}$ is z^k.*

The functions $Y_{m,k}$ are called the LAPLACE SPHERICAL FUNCTIONS.

PROOF. It follows from Lemma 3 and Theorem 1 that for each m

$$\rho(A_m) = \rho(H_m) \oplus \rho(H_{m-2}) \oplus \rho(H_{m-4}) \oplus \ldots,$$

and that $U_m = \rho(H_m)$ is a minimal SO_3-invariant subspace. Since

$$\mathbf{C}[S] = \rho(A) = \rho(A_0) + \rho(A_1) + \rho(A_2) + \ldots,$$

we have that

$$\mathbf{C}[S] = U_0 + U_1 + U_2 + \dots .$$

The representation of SO_3 in U_m is isomorphic to its representation in the space $H_m \subset A_m$. The representation of the group $SO_2 \subset SO_3$ in A_m is a sum of one-dimensional representations of the type $h(z) \mapsto z^k$ ($|k| \le m$), whose multiplicities were computed in Lemma 4. Consequently, the representation of SO_2 in H_m is also a sum of one-dimensional representations. It follows from decomposition (1) that the multiplicity of the one-dimensional representation $h(z) \mapsto z^k$ ($|k| \le m$) in the representation of SO_2 in the space H_m is equal to the difference of its multiplicities in the representations of SO_2 in the spaces A_m and A_{m-2}, which in turn is equal to 1:

$$[\frac{1}{2}(m - |k|)] - [\frac{1}{2}(m - 2 - |k|)] = 1.$$

Thus, in U_m there is a basis $(Y_{m,0}, Y_{m,\pm 1}, \dots, Y_{m,\pm m})$ such that

$$h(z)_* Y_{m,k} = z^k Y_{m,k}.$$

In particular, $\dim U_m = 2m + 1$.

Comparing dimensions we see that for $m \ne \ell$ the representations of SO_3 in U_m and U_ℓ are not isomorphic. Applying Theorem 9 of 4.7, we conclude that the subspaces U_m are mutually orthogonal, and consequently linearly independent. By the same theorem, applied now to the representation of SO_2 in U_m, the functions $Y_{m,0}, Y_{m,\pm 1}, \dots, Y_{m,\pm m}$ are pairwise orthogonal. This completes the proof of the present theorem. □

Corollary. *The functions* $Y_{m,k}$ ($m = 0, 1, 2, \dots;\ k = 0, \pm 1, \dots, \pm m$) *constitute a complete orthonormal set in* $C_2(S)$. □

9.6. Now let us find *explicit expressions for the Laplace functions.* We denote by ξ_1, ξ_2, ξ_3 the restrictions of the coordinate functions x_1, x_2, x_3 to the sphere S.

The space $U_m = \rho(H_m)$ contains a unique (up to a multiplicative constant) SO_2-invariant function, namely, $Y_{m,0}$. It is called the ZONAL SPHERICAL FUNCTION OF ORDER m. It is the restriction to S of a linear combination of polynomials of the form

$$u^p \bar{u}^p x_3^\ell = (u\bar{u})^p x_3^\ell = (x_1^2 + x_2^2)^p x_3^\ell \qquad (\text{with } 2p + \ell = m).$$

Since $\rho(x_1^2 + x_2^2) = \rho(1 - x_3^2) = 1 - \xi_3^2$, it follows that $Y_{m,0}$ is a polynomial of degree $\leq m$ in ξ_3:

$$Y_{m,0} = P_m(\xi_3).$$

The linear independence of the functions $Y_{0,0}, Y_{1,0}, \ldots, Y_{m,0}$ implies that $P_0, P_1, \ldots, P_m$ constitute a basis in the space of polynomials of degree $\leq m$.

Lemma. *$\int_{-1}^{1} P_m(t)\overline{P_\ell(t)}\, dt = 0$ for $m \neq \ell$.*

Proof. We calculate the inner product of the functions $Y_{m,0}$ and $Y_{\ell,0}$. To this end we note that the area of the infinitesimally thin belt on the sphere S specified by the inequalities $t \leq \xi_3 \leq t + dt$ equals $2\pi\, dt$. Consequently, the integral of any function of the form $P(\xi_3)$ over S equals $2\pi \int_{-1}^{1} P(t)\, dt$. In particular,

$$(Y_{m,0}, Y_{\ell,0}) = 2\pi \int_{-1}^{1} P_m(t)\overline{P_\ell(t)}\, dt$$

which, in view of the orthogonality of the functions $Y_{m,0}$ and $Y_{\ell,0}$, proves the lemma. □

We define an inner product in the space of polynomials of one variable by the rule

$$(P, Q) = \int_{-1}^{1} P(t)\overline{Q(t)}\, dt.$$

Then the lemma asserts that the polynomials $P_0, P_1, P_2, \ldots$ are mutually orthogonal. Since $(P_0, P_1, \ldots, P_{m-1})$ is a basis of the space of polynomials of degree $< m$, it follows that P_m is a polynomial of degree m that is orthogonal to all polynomials of degree $< m$. As such, it is uniquely determined up to a multiplicative constant. It is called the Legendre polynomial of degree m. We have thus established

Theorem 3. *$Y_{m,0} = P_m(\xi_3)$, where P_m is the Legendre polynomial of degree m.* □

Here are explicit expressions for a few of the first Legendre polynomials:

$$P_0(t) = 1, \qquad P_1(t) = t,$$
$$P_2(t) = 3t^2 - 1, \qquad P_3(t) = 5t^3 - 3t,$$
$$P_4(t) = 35t^4 - 30t^2 + 3, \qquad P_5(t) = 63t^5 - 70t^3 + 15t.$$

As regards the remaining spherical functions, they can be expressed through Legendre polynomials as follows:

$$(3)\qquad \begin{aligned} Y_{m,k} &= (\xi_1 - i\xi_2)^k P_m^{(k)}(\xi_3), \qquad (k > 0), \\ Y_{m,-k} &= (\xi_1 + i\xi_2)^k P_m^{(k)}(\xi_3), \qquad (k > 0). \end{aligned}$$

We are going to see this in Section 11, where we will also show that the representation of SO_3 in the space U_m is isomorphic to the representation Ψ_{2m} constructed in 7.4.

Questions and Exercises

1. Suppose the compact group G acts transitively on a topological space X. Let H be the isotropy subgroup of the point $o \in X$. Show that in every finite-dimensional nonzero G-invariant subspace of continuous functions on X there exists a nonzero H-invariant function.

2. Under the assumptions of the preceding exercise, suppose that $T: G \to \mathrm{GL}(V)$ is a finite-dimensional irreducible representation of G. Prove that if V contains a nonzero H-invariant vector, then T is isomorphic to a representation of G in a space of continuous functions on X. (*Hint:* Establish first the existence of a nonzero H-invariant linear function $f \in V'$; then consider the map that assigns to each vector $v \in V$ the continuous function f_v on X defined by the formula $f_v(go) = f(g^{-1}v)$.)

Work out Exercises 3–6 without resorting to formulas (3).

3. Show that $Y_{m,k}(o) = 0$ if $k \neq 0$.

4. Find explicit expressions for the functions $Y_{m,k}$ for $m = 1, 2$.

5. Show that $Y_{m,m} = (\xi_1 - i\xi_2)^m$.

6. Show that $\overline{U}_m = U_m$. Deduce from this that $\overline{Y}_{m,k} = Y_{m,-k}$ (where the bar denotes complex conjugation).

7. Prove the following formula of Rodrigues:

$$P_m(t) = \frac{d^m}{dt^m}\left[(t^2 - 1)^m\right].$$

IV. Representations of Lie Groups

In this chapter we deal with *differentiable* functions and maps. Differentiability will always be understood as infinite differentiability, i.e., the existence of partial derivatives of arbitrarily high order.

All vector spaces encountered below are assumed to be finite-dimensional, except for certain function spaces which are clearly infinite-dimensional.

10. General Properties of Homomorphisms and Representations of Lie Groups

10.1. A LIE GROUP is, by definition, a group endowed with a differentiable structure (i.e., differentiable manifold structure) such that the group operations

$$x \mapsto x^{-1}, \qquad (x, y) \mapsto xy$$

are differentiable maps (cf. the definition of a topological group, 2.4.) Depending on whether one considers real or complex manifolds, one speaks about REAL or COMPLEX Lie groups.

Any Lie group is a topological group.

Examples of Lie groups.

1. Any group endowed with the discrete topology and the trivial structure of a zero-dimensional manifold.

2. The additive group of any n-dimensional real or complex vector space V.

3. The group $\mathrm{GL}(V)$, with the differentiable structure that it gets as an open subset of the vector space $\mathrm{L}(V)$. That is, one declares differentiable the differentiable functions of the matrix elements.

Every complex Lie group can be regarded as a real one (of twice the dimension) upon taking as real coordinates the real and imaginary parts of the complex coordinates.

E.B. Vinberg, *Linear Representations of Groups*, Modern Birkhäuser Classics,
DOI 10.1007/978-3-0348-0063-1_5, © Birkhäuser Verlag 1989

10.2. A subgroup H of the Lie group G is said to be REGULAR, or a LIE SUBGROUP if it is a submanifold, i.e., if it can be given, in a neighborhood of any of its points h, by a system of equations of the form

$$F_i(x_1, \dots, x_n) = 0 \qquad (i = 1, \dots, m), \tag{1}$$

where $x_1, \dots, x_n$ are local coordinates on G in the neighborhood of h, the functions $F_1, \dots, F_m$ are differentiable, and

$$\operatorname{rank} \left[\frac{\partial F_i}{\partial x_j} \right] \bigg|_h = m. \tag{2}$$

A regular subgroup endowed with the canonical induced structure of an $(n-m)$-dimensional differentiable manifold is a Lie group.

Lemma. *Suppose that the subgroup H of the Lie group G can be given, in some neighborhood of the identity, by a system of equations (1) satisfying condition (2). Then H is a regular subgroup.*

PROOF. The subset $H \subset G$ is invariant under translations:

$$l(h) \colon x \mapsto hx \qquad (x \in G),$$

which are diffeomorphisms of the manifold G. We use $l(h)$ to carry the coordinate system in the neighborhood of the identity in G whose existence is asserted in the hypothesis over to a neighborhood of the point h. Then in the resulting coordinate system around h, the subgroup H is given by the same equations that defined it in the neighborhood of the identity. □

Examples.

1. $\mathrm{SL}_n(K)$ (with $K = \mathbf{R}$ or $\mathbf{C}$) is a regular subgroup of the Lie group $\mathrm{GL}_n(K)$. In fact, it is given by the single equation

$$\det X = 1 \qquad (X = [x_{ij}] \in \mathrm{GL}_n(K)),$$

and

$$\frac{\partial \det X}{\partial x_{11}} \bigg|_{X=E} = 1 \neq 0.$$

Thus, $\mathrm{SL}_n(K)$ is an (n^2-1)-dimensional Lie group.

2. *The orthogonal group* O_n is singled out in $\mathrm{GL}_n(\mathbf{R})$ by the equations

$$F_{ij}(X) = \sum_k x_{ik} x_{jk} = \delta_{ij},$$

where one can assume that $i \leq j$, so that there are $\frac{1}{2}n(n+1)$ equations. We have that

$$\left.\frac{\partial F_{ij}(X)}{\partial x_{st}}\right|_{X=E} = \begin{cases} 2 \text{ or } 1, & \text{if } \{i,j\} = \{s,t\}, \\ 0, & \text{if } \{i,j\} \neq \{s,t\}. \end{cases}$$

It follows that the minor of order $\frac{1}{2}n(n+1)$ of the Jacobian matrix $\left[\frac{\partial F_{ij}(X)}{\partial x_{st}}\right]$ corresponding to the variables x_{st}, with $s \leq t$, does not vanish at the point E. Thus, O_n is a regular subgroup of the Lie group $\mathrm{GL}_n(\mathbf{R})$. It has dimension

$$n^2 - \frac{1}{2}n(n+1) = \frac{1}{2}n(n-1).$$

3. The group SO_n coincides in a neighborhood of the identity with O_n (as a subgroup of $\mathrm{GL}_n(\mathbf{R})$). In fact, the determinant of an orthogonal matrix can assume the values ± 1, but if the matrix is close to the identity matrix, its determinant cannot be equal to -1. Hence, SO_n is a regular subgroup of the Lie group $\mathrm{GL}_n(\mathbf{R})$. It has the same dimension as O_n. One can show that SO_n is a (connected) component of the group O_n.

4. The *unitary group* U_n is singled out in $\mathrm{GL}_n(\mathbf{C})$ by the equations

$$\sum_k x_{ik}\bar{x}_{jk} = \delta_{ij} \qquad (X = [x_{ij}] \in \mathrm{GL}_n(\mathbf{C})),$$

where one can assume that $i \leq j$. If $\mathrm{GL}_n(\mathbf{C})$ is regarded as a real Lie group, the left-hand sides of these equations are differentiable functions. On separating the real and imaginary parts one obtains n^2 equations. Proceeding by analogy with Example 2, one can show that the rank of the Jacobian matrix of interest at the point $X = E$ is equal to the number of equations. Consequently, U_n is a regular subgroup of the real Lie group $\mathrm{GL}_n(\mathbf{C})$. Its dimension is equal to

$$2n^2 - n = n^2.$$

10.3. A HOMOMORPHISM OF THE LIE GROUP G INTO THE LIE GROUP H is a group homomorphism of G into H which is also a differentiable map. Lie group ISOMORPHISMS and AUTOMORPHISMS are defined in an analogous manner.

A LINEAR REPRESENTATION OF THE LIE GROUP G is, by definition, a homomorphism of G into the Lie group $\mathrm{GL}(V)$, where V is a real or complex vector space. The following variants can then be distinguished:

real linear representations of a *real* Lie group;

complex linear representations of a *real* Lie group (the space V is complex, but $\mathrm{GL}(V)$ is regarded as a real Lie group); and

complex linear representations of a *complex* Lie group.

The differentiability of a linear representation T of the Lie group G means that the matrix elements of the operators $T(g)$ are differentiable functions of g.

Examples.

1. Every real or complex linear representation of an arbitrary group G is simultaneously a linear representation of the discrete Lie group G.

2. The map $t \mapsto \mathrm{e}^{it}$ ($t \in \mathbf{R}$) is a Lie group homomorphism of $\mathbf{R}$ onto $\mathbf{T} = \{ z \in \mathbf{C} \mid |z| = 1 \}$.

3. The map $t \mapsto \mathrm{e}^{tA}$ ($t \in \mathbf{R}$, $A \in \mathrm{L}_n(\mathbf{R})$ or $\mathrm{L}_n(\mathbf{C})$) is a (real or, respectively, complex) representation of the Lie group $\mathbf{R}$. The map $t \mapsto \mathrm{e}^{tA}$ ($t \in \mathbf{C}$, $A \in \mathrm{L}_n(\mathbf{C})$) is a linear representation of the complex Lie group $\mathbf{C}$.

4. Let V be a real or complex vector space. Then all linear representations of $\mathrm{GL}(V)$ considered in 1.6 are differentiable, i.e., are linear representations of the Lie group $\mathrm{GL}(V)$. For instance, the matrix elements of the representation Φ (Example 3, 1.6) were calculated in 2.4. It is obvious that they are differentiable and not merely continuous, which means that Φ is a differentiable map.

5. It follows from the differentiability of the group operations in a Lie group G that every inner automorphism

$$a(g): x \mapsto gxg^{-1} \qquad (g, x \in G)$$

is differentiable, i.e., it is a Lie group automorphism of G.

6. The restriction of a representation of a Lie group G to a regular subgroup $H \subset G$ is a representation of the Lie group H.

10.4. The Tangent Map of a Homomorphism. We shall denote by $D_x X$ the tangent space to the manifold X at the point $x \in X$, and by $d_x\phi\colon D_x X \to D_{\phi(x)} Y$ the tangent map (the differential) of the differentiable map $\phi\colon X \to Y$ at x. Whenever the argument of the map $d_x\phi$ (an element of $D_x X$) is indicated, the subscript x in the notation $d_x\phi$ will be omitted.

Now let $\phi\colon G \to H$ be a Lie group homomorphism. Then $\phi(e) = e$, and there is defined the linear map

$$d_e\phi\colon D_e G \to D_e H.$$

We call $d_e\phi$ the TANGENT MAP of the homomorphism ϕ.

Theorem 1. *Any homomorphism of a connected Lie group G into an arbitrary Lie group H is uniquely determined by its tangent map.*

This is the first genuine theorem of the theory of Lie groups. We precede its proof with some notation and a lemma.

Consider the tangent map of the left translation $l(g)$ on G at the point e. It is an isomorphism of the tangent spaces:

$$d_e l(g)\colon D_eG \to D_gG. \tag{3}$$

We shall denote the image of $\xi \in D_eG$ under this isomorphism simply by $g\xi$. (If $G = \mathrm{GL}(V)$, then $g\xi$ can be understood as the usual product; see 10.5 below.)

Lemma. *Let $\phi\colon G \to H$ be a Lie group homomorphism. Then for any $g \in G$ and any $\xi \in D_eG$,*

$$d\phi(g\xi) = \phi(g)d\phi(\xi). \tag{4}$$

Proof. Since ϕ is a homomorphism we have the following commutative diagram of maps:

$$\begin{array}{ccc} G & \xrightarrow{l(g)} & G \\ {\scriptstyle\phi}\downarrow & & \downarrow{\scriptstyle\phi} \\ H & \xrightarrow{l(\phi(g))} & H \end{array}$$

(In fact, $l(\phi(g))\phi(x) = \phi(g)\phi(x) = \phi(gx) = \phi(l(g)x)$ for all $x \in G$.) It yields the corresponding commutative diagram of tangent maps

$$\begin{array}{ccc} D_eG & \xrightarrow{d_el(g)} & D_gG \\ {\scriptstyle d_e\phi}\downarrow & & \downarrow{\scriptstyle d_g\phi} \\ D_eH & \xrightarrow{d_el(\phi(g))} & D_{\phi(g)}H \end{array}$$

We get equality (4) by just writing out what the commutativity of this diagram means for an element $\xi \in D_eG$. □

Proof of Theorem 1. Let $d_e\phi$ be the tangent map of the homomorphism $\phi\colon G \to H$. Let us show how one can recover ϕ from $d_e\phi$.

In order to find $\phi(g)$, we connect the identity e of G with the point g by a differentiable curve $g(t)$, $0 \le t \le 1$:

$$g(0) = e, \qquad g(1) = g.$$

For each t we use the isomorphism (3) to write the tangent vector $g'(t) \in D_{g(t)}G$ in the form

(5) $g'(t) = g(t)\xi(t)$, with $\xi(t) \in D_eG$.

By the lemma,

(6) $d\phi(g'(t)) = \phi(g(t))d\phi(\xi(t))$.

Set

$$\phi(g(t)) = h(t) \quad \text{and} \quad d\phi(\xi(t)) = \eta(t).$$

Then, by the definition of the tangent map,

$$d\phi(g'(t)) = h'(t),$$

and (6) can be reexpressed as

(7) $h'(t) = h(t)\eta(t)$.

If the tangent map $d_e\phi$ is known, so is $\eta(t)$; hence (7) can be regarded as a system of differential equations for the coordinates of $h(t)$. The solution of this system is uniquely determined by the initial condition $h(0) = e$. In particular, the point $h(1) = \phi(g)$ is uniquely determined, as we needed to show. □

We wish to emphasize that Theorem 1 is concerned with the *uniqueness* of the Lie group homomorphism with a given tangent map and not with its *existence*.

Example. We use Theorem 1 to describe all automorphisms of the Lie group $\mathbf{R}$. The tangent map of an arbitrary automorphism ϕ of $\mathbf{R}$ is an invertible linear map of the one-dimensional space $D_0\mathbf{R}$, i.e., reduces to multiplication by some $c \in \mathbf{R}$, $c \neq 0$. Now, notice that the transformation $x \mapsto cx$ is an automorphism of the group $\mathbf{R}$, and that its tangent map is precisely multiplication by c. Hence there are no other automorphisms. (Resorting to the proof of Theorem 1, one could determine the form that an automorphism must have by solving a differential equation rather than producing it from thin air, as above.)

10.5. The Tangent Representation. The notion of tangent homomorphism can be applied to linear representations of Lie groups.

Let V be a real or complex vector space. The group $\mathrm{GL}(V)$ is an open subset of the vector space $\mathrm{L}(V)$. For each $\alpha \in \mathrm{GL}(V)$ (in particular, for $\alpha = \varepsilon$), the tangent space $D_\alpha\mathrm{GL}(V)$ can be canonically identified with $\mathrm{L}(V)$. By this we mean that the tangent vector to a curve passing through α is identified with the ordinary derivative of the (vector) $\mathrm{L}(V)$-valued function that specifies the curve. Under this identification the tangent space to the left translation $l(\alpha)$ becomes the operator of left multiplication by α in the algebra $\mathrm{L}(V)$.

Definition. Let $T: G \to \mathrm{GL}(V)$ be a linear representation of the Lie group G. We call the tangent map

$$d_eT: D_eG \to \mathrm{L}(V) = D_\varepsilon \mathrm{GL}(V)$$

the TANGENT REPRESENTATION for T. (Note that d_eT is not a representation of some group; as we shall see below, it is a representation of a *Lie algebra.*)

Corollary of Theorem 1. *Any linear representation of a connected Lie group is uniquely determined by its tangent representation.* □

Examples.

1. The tangent representation for the linear representation

$$F_\alpha: t \mapsto \mathrm{e}^{t\alpha} \qquad (t \in \mathbf{R}, \quad \alpha \in \mathrm{L}(V))$$

of $\mathbf{R}$ is the map

$$(8) \qquad t \mapsto t\alpha.$$

Since any linear map $\mathbf{R} \to \mathrm{L}(V)$ can be written in the form (8) for a suitable $\alpha \in \mathrm{L}(V)$, it follows that every linear representation of $\mathbf{R}$ in V is of the type F_α. It is precisely in this manner, by actually carrying out in a particular case the proof of Theorem 1 given above, that we found in 0.1 all differentiable representations of $\mathbf{R}$.

2. Let us determine the tangent representation for the *adjoint representation* Ad of the group $\mathrm{GL}(V)$ (Example 2, 1.6). Let $\alpha(t)$, $0 \le t \le 1$, be a curve in the group $\mathrm{GL}(V)$ with initial data

$$\alpha(0) = \varepsilon, \qquad \alpha'(0) = \xi \in \mathrm{L}(V).$$

For every $\eta \in \mathrm{L}(V)$ we have

$$\begin{aligned} d\mathrm{Ad}(\xi)\eta &= \frac{d}{dt}\mathrm{Ad}(\alpha(t))\eta\Big|_{t=0} \\ &= \frac{d}{dt}\alpha(t)\eta\alpha(t)^{-1}\Big|_{t=0} = \xi\eta - \eta\xi. \end{aligned}$$

(Here we used the fact that $\frac{d}{dt}\alpha(t)^{-1}\big|_{t=0} = -\xi$, which follows upon differentiating the identity $\alpha(t)\alpha(t)^{-1} = \varepsilon$ at $t = 0$.)

The tangent representation $d_e\mathrm{Ad}$ is denoted by ad. We have shown that

$$(9) \qquad \mathrm{ad}(\xi)\eta = \xi\eta - \eta\xi.$$

10.6. In the case when the given Lie group is *connected*, the properties of its linear representations are determined by the properties of the corresponding tangent representations. This holds true, in particular, for the invariant subspaces.

Let $T: G \to \mathrm{GL}(V)$ be a linear representation of the Lie group G and $d_eT: D_eG \to \mathrm{L}(V)$ the tangent representation for T. We will say that the subspace $U \subset V$ is *invariant* under the tangent representation d_eT if

$$dT(\xi)u \in U \qquad \text{for all } \xi \in D_eG \text{ and all } u \in U.$$

Theorem 2. a) *Every subspace U invariant under T is also invariant under d_eT.*

b) *If G is connected, then every subspace U invariant under d_eT is also invariant under T.*

PROOF. a) Suppose U is invariant under T. Given an arbitrary $\xi \in D_eG$ we find a curve $g(t)$, $0 \leq t \leq 1$, in G such that

$$g(0) = e, \qquad g'(0) = \xi.$$

Since, for each given $u \in U$,

$$T(g(t))u \in U \qquad \text{for all } t,$$

we have that

$$dT(\xi)u = \frac{d}{dt}T(g(t))u\Big|_{t=0} \in U.$$

Hence, U is invariant under d_eT.

b) Suppose now that G is connected and U is invariant under d_eT. To show that $T(g)U \subset U$, we join the identity element e of G with the point g by a differentiable curve $g(t)$, $0 \leq t \leq 1$:

$$g(0) = e, \qquad g(1) = g.$$

We have

$$g'(t) = g(t)\xi(t), \qquad \text{with } \xi(t) \in D_eG$$

(cf. the proof of Theorem 1). By the Lemma of 10.4, applied to the homomorphism $T: G \to \mathrm{GL}(V)$,

$$\frac{d}{dt}T(g(t)) = T(g(t))dT(\xi(t)). \tag{10}$$

Now pick a basis $(e) = (e_1, \dots, e_n)$ in V such that $U = \langle e_1, \dots, e_k \rangle$, and set

$$T_{(e)}(g(t)) = A(t) = [a_{ij}(t)]$$

and

$$dT_{(e)}(\xi(t)) = C(t) = [c_{ij}(t)].$$

By hypothesis,

$$(11) \qquad c_{ij}(t) = 0 \qquad \text{for } i > k, \quad j \le k.$$

We prove that

$$(12) \qquad a_{ij}(t) = 0 \qquad \text{for } i > k, \quad j \le k.$$

To this end we rewrite (10) in terms of matrices:

$$A'(t) = A(t)C(t).$$

Taking into account (11), this yields for $i > k$ and $j \le k$

$$(13) \qquad a'_{ij}(t) = \sum_{\ell=1}^{k} a_{i\ell}(t)c_{\ell j}(t).$$

These equalities can be regarded as a system of differential equations for the functions $a_{ij}(t)$ $(i > k,\ j \le k)$. Since $A(0) = E$, $a_{ij}(0) = 0$ for $i > k$, $j \le k$, and (12) follows from the uniqueness theorem for the solution of a system of differential equations with given initial condition. In particular, setting $t = 1$ we conclude that $T(g)U \subset U$. □

Corollary. a) *If the tangent representation d_eT for T is irreducible, then so is T.*

b) *If G is connected and the representation T is irreducible, then so is the tangent representation d_eT.* □

This corollary will be used in Section 11.

10.7. The Tangent Algebra. The tangent space D_eG to a Lie group G does not, by itself, carry any information on the multiplication operation in G. However, the latter can be used to define, in a certain canonical manner, a binary operation in the space D_eG called *commutation*, which turns D_eG into an algebra. If G is connected, then the structure of this algebra determines to a considerable extent the structure of the group G itself.

The COMMUTATOR of the elements $\xi, \eta \in D_eG$, denoted by $[\xi, \eta]$, is defined by the formula

$$(14) \qquad [\xi, \eta] = \frac{\partial^2}{\partial t \partial s}(g(t), h(s))\Big|_{t=s=0},$$

where $g(t)$, $0 \le t \le 1$, and $h(s)$, $0 \le s \le 1$, are curves in G satisfying the conditions $g(0) = h(0) = e$, $g'(0) = \xi$, $h'(0) = \eta$, and $(g,h) = ghg^{-1}h^{-1}$ denotes the commutator of the elements g, h in the group G.

The partial derivative in (14) can be defined coordinate-wise but is independent of the choice of the coordinate system, as it admits the following alternative interpretation. Set

$$f(t,s) = (g(t), h(s)).$$

Then $f(t,0) = e$ and

$$\xi(t) = \frac{\partial}{\partial s} f(t,s)\Big|_{s=0} \in D_e G$$

is the tangent vector to the curve $s \mapsto f(t,s)$. Now $\frac{\partial^2}{\partial t \partial s} f(t,s)\big|_{t=s=0}$ is seen to be the tangent vector to the curve $\xi(t)$ in the space $D_e G$.

It follows from the definition of a Lie group that there exist differentiable functions f_i of $2n$ variables $x_1, \dots, x_n, y_1, \dots, y_n$ such that the i-th coordinate of the commutator of the elements of G with coordinates $(x_1, \dots, x_n)$ and $(y_1, \dots, y_n)$ is $f_i(x_1, \dots, x_n, y_1, \dots, y_n)$. Regarding the f_i as known functions, we obtain the following explicit expression for the coordinates of the commutator $[\xi, \eta]$:

$$(15) \qquad [\xi, \eta]_i = \sum_{j,k} \frac{\partial^2 f_i}{\partial x_j \partial y_k}\Big|_{(e,e)} \xi_j \eta_k.$$

This expression shows, first, that the commutator $[\xi, \eta]$ does not depend on the choice of the curves $g(t)$ and $h(s)$ and, second, that it is linear in each of its factors.

Definition. We call the tangent space $D_e G$, endowed with the commutator defined by the rule (14) (or (15)), TANGENT ALGEBRA of the Lie group G.

The tangent algebra is usually denoted by the lower-case Gothic letter (or letters) that correspond to the Roman letter(s) denoting the Lie group. Thus, the tangent algebra of the Lie group G is denoted by $\mathfrak{g}$, that of $\mathrm{SL}_n(\mathbf{C})$ by $\mathfrak{sl}_n(\mathbf{C})$, and so on.

Examples.

1. If the Lie group G is commutative, then $[\xi, \eta] = 0$ for all $\xi, \eta \in \mathfrak{g}$.

2. Let us determine the tangent algebra of $\mathrm{GL}(V)$. Given arbitrary $\xi, \eta \in \mathrm{L}(V) = D_\varepsilon \mathrm{GL}(V)$, we pick curves $\alpha(t), \beta(s) \in \mathrm{GL}(V)$ such that

$$\alpha(0) = \beta(0) = \varepsilon, \qquad \alpha'(0) = \xi, \qquad \beta'(0) = \eta.$$

Then (see Example 2, 10.5),

$$\frac{d}{dt}\alpha(t)^{-1}\Big|_{t=0} = -\xi, \qquad \frac{d}{ds}\beta(s)^{-1}\Big|_{s=0} = -\eta,$$

$$\frac{\partial}{\partial s}\alpha(t)\beta(s)\alpha(t)^{-1}\beta(s)^{-1}\Big|_{s=0} = \alpha(t)\eta\alpha(t)^{-1} - \eta,$$

and

$$\frac{\partial}{\partial t}(\alpha(t)\eta\alpha(t)^{-1} - \eta)\Big|_{t=0} = \xi\eta - \eta\xi.$$

Thus,

$$[\xi, \eta] = \xi\eta - \eta\xi. \tag{16}$$

It is readily verified that the commutation operation in $\mathrm{L}(V)$ possesses the following properties:

$$[\xi, \eta] + [\eta, \xi] = 0, \qquad (\textit{skew-symmetry}, \text{ or } \textit{anti-commutativity})$$

and

$$[\xi, [\eta, \zeta]] + [\eta, [\zeta, \xi]] + [\zeta, [\xi, \eta]] = 0 \qquad (\text{the } \textit{Jacobi identity}).$$

Any algebra in which these two identities hold is called a LIE ALGEBRA. One can show that *the tangent algebra of any Lie group is a Lie algebra.* We shall not need this fact, however.

It is obvious that *the tangent algebra* $\mathfrak{h}$ *of a regular subgroup* H *of the Lie group* G *is a subalgebra of* $\mathfrak{g}$. In particular, if H is a regular subgroup of the Lie group $\mathrm{GL}(V)$, then $\mathfrak{h}$ is a subspace of $\mathrm{L}(V)$, and the commutator $[\xi, \eta]$ of $\xi, \eta \in \mathfrak{h}$ is computed according to formula (16).

We remark that if the subgroup H of the Lie group G is given in a neighborhood of the identity by a system of equations (1) with property (2), then the subspace D_eH is singled out in D_eG by the linear equations

$$\sum_j \frac{\partial F_i}{\partial x_j}\Big|_e \xi_j = 0,$$

where $(\xi_1, \ldots, \xi_n)$ is the coordinate system on the tangent space D_eG corresponding to the coordinate system $(x_1, \ldots, x_n)$ on G.

3. O_n is a regular subgroup of $\mathrm{GL}_n(\mathbf{R})$ (see Example 2, 10.2). The system of equations defining O_n is equivalent to the matrix equation

$$AA' = E.$$

Differentiating the latter with respect to A at $A = E$, we obtain the equation of the tangent space $D_E\mathrm{O}_n$:

$$X + X' = 0.$$

Thus, the tangent algebra of the Lie group O_n is the algebra of the skew-symmetric real matrices with the commutation operation

$$[X, Y] = XY - YX. \tag{17}$$

In particular, we see that the commutator of two skew-symmetric matrices is again a skew-symmetric matrix (which, of course, can also be verified directly).

In a similar fashion one proves that the tangent algebra of the Lie group U_n is the algebra of skew-Hermitian complex matrices with the commutator (17).

10.8. In 10.4 we saw that a homomorphism of connected Lie groups is uniquely determined by its tangent map. We are led to ask which linear maps are tangent maps of Lie group homomorphisms. A partial answer is provided by

Theorem 3. *The tangent map of a Lie group homomorphism is a homomorphism of the corresponding tangent algebras.*

PROOF. Let $\phi: G \to H$ be a Lie group homomorphism. Given $\xi, \eta \in \mathfrak{g}$, choose curves $g(t), h(s)$ as in the definition of the commutator. By the definition of the tangent map,

$$\begin{aligned} d\phi\left(\frac{\partial}{\partial s}(g(t), h(s))\Big|_{s=0}\right) &= \frac{\partial}{\partial s}\phi((g(t), h(s)))\Big|_{s=0} \\ &= \frac{\partial}{\partial s}(\phi(g(t)), \phi(h(s)))\Big|_{s=0}. \end{aligned}$$

Since the map $d_e\phi$: $\mathfrak{g} \to \mathfrak{h}$ is linear and continuous, it commutes with differentiation. Consequently,

$$\begin{aligned} d\phi\left(\frac{\partial^2}{\partial t \partial s}(g(t), h(s))\Big|_{t=s=0}\right) &= \frac{\partial}{\partial t} d\phi\left(\frac{\partial}{\partial s}(g(t), h(s))\Big|_{s=0}\right)\Big|_{t=0} \\ &= \frac{\partial^2}{\partial t \partial s}(\phi(g(t)), \phi(h(s))\Big|_{t=s=0}, \end{aligned}$$

i.e.,

$$d\phi([\xi, \eta]) = [d\phi(\xi), d\phi(\eta)],$$

as we needed to show. □

Corollary. *Let $T: G \to \mathrm{GL}(V)$ be a linear representation of the Lie group G. Then*

$$dT([\xi, \eta]) = dT(\xi)dT(\eta) - dT(\eta)dT(\xi)$$

for every $\xi, \eta \in \mathfrak{g}$. □

For an arbitrary Lie algebra $\mathfrak{a}$, a linear map $\tau: \mathfrak{a} \to \mathrm{L}(V)$ satisfying the condition:

$$\tau([\xi, \eta]) = \tau(\xi)\tau(\eta) - \tau(\eta)\tau(\xi) \qquad \text{for all } \xi, \eta \in \mathfrak{a},$$

is called a LINEAR REPRESENTATION OF $\mathfrak{a}$. Thus, the tangent map d_eT of any representation T of the Lie group G is a linear representation of the tangent algebra $\mathfrak{g}$.

10.9. The Adjoint Representation. For each $g \in G$ we let $a(g)$ denote the inner automorphism of G defined by g:

$$a(g)x = gxg^{-1} \qquad (x \in G).$$

Suppose G is a Lie group. Then we can consider the tangent map $d_ea(g) \in \mathrm{GL}(\mathfrak{g})$, denoted $\mathrm{Ad}(g)$. Since $a(g_1g_2) = a(g_1)a(g_2)$, we have that $\mathrm{Ad}(g_1g_2) = \mathrm{Ad}(g_1)\mathrm{Ad}(g_2)$. Moreover, it follows from the definition of a Lie group that $\mathrm{Ad}(g)$ depends differentiably on g. Hence, the map

$$\mathrm{Ad}: g \mapsto \mathrm{Ad}(g)$$

is a linear representation of G in the vector space $\mathfrak{g}$. It is called the ADJOINT REPRESENTATION OF THE LIE GROUP G.

When $G = \mathrm{GL}(V)$, this general definition of the adjoint action is in agreement with the one given in Example 2 of 1.6. In fact, in this case $a(g)x$ depends linearly on x, and so $\mathrm{Ad}(g)$ is given by the same formula as $a(g)$. Specifically, for $\xi \in \mathfrak{g} = \mathrm{L}(V)$,

$$\mathrm{Ad}(g)\xi = g\xi g^{-1}. \tag{18}$$

The same formula is, needless to say, valid for every regular subgroup $G \subset \mathrm{GL}(V)$.

The tangent representation for Ad is called the ADJOINT REPRESENTATION OF THE TANGENT ALGEBRA $\mathfrak{g}$ and is denoted by ad. In 10.5 (Example 2) we showed that, for $G = \mathrm{GL}(V)$,

$$\mathrm{ad}(\xi)\eta = [\xi, \eta] \qquad (\xi, \eta \in \mathfrak{g}). \tag{19}$$

In exactly the same manner one can prove that this formula holds for any regular subgroup $G \subset \mathrm{GL}(V)$. Actually, one can show that (19) is valid for any Lie group G.

By the Corollary to Theorem 3,

$$\mathrm{ad}([\xi,\eta]) = \mathrm{ad}(\xi)\mathrm{ad}(\eta) - \mathrm{ad}(\eta)\mathrm{ad}(\xi).$$

In view of (19), this means that

$$[[\xi,\eta],\zeta] = [\xi,[\eta,\zeta]] - [\eta,[\xi,\zeta]] \tag{20}$$

for all $\xi,\eta,\zeta \in \mathfrak{g}$. Assuming that the skew-symmetry of the commutation operation is a known fact, (20) is equivalent to the Jacobi identity. This is in fact one of the ways of proving the latter.

10.10. The next theorem describes the behavior of the adjoint representation under a homomorphism.

Theorem 4. *Let $\phi\colon G \to H$ be a Lie group homomorphism. Then*

$$d\phi(\mathrm{Ad}(g)\xi) = \mathrm{Ad}(\phi(g))d\phi(\xi) \tag{21}$$

for all $g \in G$ and all $\xi \in \mathfrak{g}$.

PROOF. Let $x(t)$ be a curve in G tangent to the vector ξ at $t = 0$. Then $\mathrm{Ad}(g)\xi$ is the tangent vector (also at $t = 0$) to the curve $gx(t)g^{-1}$, while $d\phi(\mathrm{Ad}(g)\xi)$ is the tangent vector to the curve

$$\phi(gx(t)g^{-1}) = \phi(g)\phi(x(t))\phi(g)^{-1}.$$

This yields (21), because $d\phi(\xi)$ is the tangent vector to the curve $\phi(x(t))$. □

Corollary. *Let $T\colon G \to \mathrm{GL}(V)$ be a linear representation of the Lie group G. Then*

$$dT(\mathrm{Ad}(g)\xi) = T(g)dT(\xi)T(g)^{-1}$$

for all $g \in G$ and all $\xi \in \mathfrak{g}$. □

10.11. Velocity Fields. Suppose that there is given a differentiable action s of the Lie group G on a differentiable manifold X. (The differentiability of s means that the coordinates of the point gx are differentiable functions of the coordinates of g and x). Then with each element $\xi \in \mathfrak{g}$ we can associate a vector field on X, a "velocity field" of the action s.

Specifically, let $g(t)$ be a curve in G tangent to ξ at $t = 0$. For each $x \in X$ we put

$$\xi x = \frac{d}{dt}g(t)x\Big|_{t=0} \in D_xX. \tag{22}$$

Obviously, the coordinates of the vector ξx are linear functions of the first derivatives of the coordinates of $g(t)$ at $t = 0$, i.e., of the coordinates of ξ. If s is a linear representation we have

$$\xi x = ds(\xi)x,$$

where ds denotes the tangent representation. In the general case we shall use $ds(\xi)$ to denote the vector field $x \mapsto \xi x$. It is natural to regard the linear map $\xi \mapsto ds(\xi)$ as the "tangent action" associated with s. We call the vector field $ds(\xi)$ the VELOCITY FIELD (or the INFINITESIMAL GENERATOR) OF THE ACTION s corresponding to the element $\xi \in \mathfrak{g}$.

For any differentiable function f we have that

$$(23) \qquad \frac{d}{dt}(s_*(g(t))f)(x)\Big|_{t=0} = \frac{d}{dt}f(g(t)^{-1}x)\Big|_{t=0} = -\partial_{\xi x}f(x),$$

where $\partial_{\xi x}$ denotes the operator of differentiation in the direction of the vector ξx. From (23) we derive the following important result.

Proposition. *Let W be a finite-dimensional space of differentiable functions on X which is invariant under the representation s_*, and let $T = (s_*)_W$. Then*

$$dT(\xi)f = -\partial_{ds(\xi)}f$$

for all $\xi \in \mathfrak{g}$ and all $f \in W$.

PROOF. Let $g(t)$ be a curve in G tangent to ξ for $t = 0$. By the definition of the tangent representation,

$$dT(\xi) = \frac{d}{dt}(T(g(t)))\Big|_{t=0}.$$

For $f \in W$ we have

$$dT(\xi)f = \frac{d}{dt}T(g(t))f\Big|_{t=0} = \frac{d}{dt}s_*(g(t))f\Big|_{t=0}$$

and the needed equality now follows from (23).

Remark. Let s_* denote the representation of G in the space of all differentiable functions on X. It is natural to define the tangent representation for s_* by the rule

$$ds_*(\xi)f = \frac{d}{dt}s_*(g(t))f\Big|_{t=0}.$$

(Notice that the definition of a tangent representation given in 10.5 does not apply here since the representation s_* is infinite-dimensional.) Then, by the foregoing discussion,

$$(24) \qquad ds_*(\xi) = -\partial_{ds(\xi)}.$$

Questions and Exercises

1. Prove that U_n and $\mathrm{SU}_n = \{ A \in \mathrm{U}_n \mid \det A = 1 \}$ are regular subgroups of $\mathrm{GL}_n(\mathbf{C})$, regarded as a real Lie group.

2. Prove that the adjoint representation of $\mathrm{GL}(V)$ is differentiable.

3. Use Theorem 1 to find all Lie group homomorphisms of $\mathbf{R}$ into $\mathbf{T}$.

4. Use the Corollary to Theorem 1 to find all irreducible complex representations of the Lie group $\mathbf{R}^n$ (with addition of vectors as the group operation).

5. Find the tangent representation for the linear representation Φ of $\mathrm{GL}(V)$ in the space of bilinear functions (Example 3, 1.6).

6. Let $T: G \to \mathrm{GL}(V)$ be a linear representation of the Lie group G, (e) a basis of the space V, and (ε) the dual basis of V'. Prove that

$$dT'_{(\varepsilon)}(\xi) = -(dT_{(e)}(\xi))' \qquad \text{for all } \xi \in \mathfrak{g}.$$

7. Find the tangent algebra for the following Lie groups:

a) $\mathrm{SL}_n(\mathbf{R})$;

b) SU_n;

c) the group of real nonsingular triangular matrices of order n.

8. Prove the skew-symmetry of the commutation operation in the tangent algebra of an arbitrary Lie group.

9.* Establish formula (19) for an arbitrary Lie group.

10. Use formula (19) to show that if in the tangent algebra of the connected Lie group G the commutator of any pair of elements is equal to zero, then G is commutative.

11.* Deduce that the Jacobi identity holds in the tangent algebra of a Lie group G from the fact that the operators $\mathrm{Ad}(g)$, $g \in G$, are automorphisms of the tangent algebra.

12.* Find the tangent representation for the representation s_* of $\mathrm{GL}_2(\mathbf{R})$ in the space of functions on the projective line $\widehat{\mathbf{R}}$ (see Exercise 12, Section 0).

11. Representations of SU_2 and SO_3

In this section we use methods of the theory of Lie groups to describe all linear representations of the groups SU_2 and SO_3.

11.1. SU_2 and SO_3 are compact topological groups. At the same time, they are three-dimensional real Lie groups (see 10.2).

As its explicit definition shows, the homomorphism $P: SU_2 \to SO_3$ exhibited in 7.2 is differentiable, and not merely continuous. The same holds true for the linear representations Φ_n of SU_2 constructed in 7.4.

The representations of SO_3 in the spaces U_m of spherical functions are also differentiable. In fact, they are isomorphic to subrepresentations of SO_3 in spaces of homogeneous polynomials in the variables x_1, x_2, x_3, and it is readily verified that the matrix elements of the latter are polynomials in the matrix elements of the identity representation of SO_3.

[Theorem 2 of 8.3 shows that the matrix elements of any continuous linear representation of a compact linear group G belong to the algebra $\mathbf{C}[G]$. If G is a compact linear Lie group the elements of $\mathbf{C}[G]$ are differentiable functions on G. Hence, *every continuous linear representation of a compact linear Lie group is differentiable.* This is valid, in particular, for SU_2 and SO_3. As a matter of fact, *every compact linear group is a linear Lie group* (i.e., a regular subgroup of the general linear group).]

The tangent algebra $\mathfrak{su}_2$ of SU_2 consists of the traceless skew-Hermitian matrices of second order, i.e., matrices of the form

$$\begin{pmatrix} ix_1 & -x_2 + ix_3 \\ x_2 + ix_3 & -ix_1 \end{pmatrix} \qquad (x_1, x_2, x_3 \in \mathbf{R}).$$

In the notation of 7.2, $\mathfrak{su}_2 = i\mathbf{E}$. The linear representation P of SU_2 in the space $\mathbf{E}$, considered in 7.2, is isomorphic to its adjoint representation. (The isomorphism is realized through multiplication by i.)

11.2. In order to describe the (differentiable) *linear representations of* SU_2 we use the results of the preceding section, which permit us to reduce our task to that of describing the linear representations of the Lie algebra $\mathfrak{su}_2$. This in turn is facilitated by passing to the *complexification* of the algebra $\mathfrak{su}_2$. We discuss this in more detail.

Let $\mathfrak{a}$ be a real Lie algebra, and let $(e_1, \ldots, e_n)$ be a vector space basis of $\mathfrak{a}$. Then

(1) $$[e_i, e_j] = \sum_k c_{ijk} e_k \qquad (\text{with } c_{ijk} \in \mathbf{R}).$$

Consider the complexification $\mathfrak{a}_{\mathbf{C}}$ of the vector space $\mathfrak{a}$ (see 3.5 and 3.6). Define in $\mathfrak{a}_{\mathbf{C}}$ a bilinear commutation operation in such a manner that the basis vectors $e_1, \ldots, e_n$ satisfy relations (1). Then the skew-symmetry and Jacobi identities hold for the basis vectors in $\mathfrak{a}_{\mathbf{C}}$ because they hold in $\mathfrak{a}$. These identities will then hold for arbitrary vectors in $\mathfrak{a}_{\mathbf{C}}$, thanks to their linearity in all arguments. The complex Lie algebra $\mathfrak{a}_{\mathbf{C}}$ obtained in this manner is called the COMPLEXIFICATION OF THE LIE ALGEBRA $\mathfrak{a}$.

It is clear that the commutator of any two elements in $\mathfrak{a}$ coincides with their commutator in $\mathfrak{a}_{\mathbf{C}}$. This implies that the commutation operation in $\mathfrak{a}_{\mathbf{C}}$ does not depend on the choice of the basis $(e_1, \ldots, e_n)$.

Similar arguments show that *every complex linear representation of the Lie algebra $\mathfrak{a}$ extends (uniquely) to a representation of the complex Lie algebra $\mathfrak{a}_{\mathbf{C}}$ in the same space.* (By definition, a complex linear representation of a real Lie algebra $\mathfrak{a}$ is any homomorphism of $\mathfrak{a}$ into the Lie algebra of all linear operators in a complex vector space, regarded as a real algebra.) Moreover, the original and extended representations share the same invariant subspaces.

Each element of the algebra $\mathfrak{a}_{\mathbf{C}}$ is uniquely expressible as $\xi + i\eta$ with $\xi, \eta \in \mathfrak{a}$. The extension of any given complex linear representation τ of $\mathfrak{a}$ to $\mathfrak{a}_{\mathbf{C}}$ can be described by the rule

$$(2) \qquad \tau(\xi + i\eta) = \tau(\xi) + i\tau(\eta).$$

The complexification of $\mathfrak{su}_2$ is the Lie algebra $\mathfrak{sl}_2(\mathbf{C})$ of all complex traceless matrices of order two (the tangent algebra of the group $\mathrm{SL}_2(\mathbf{C})$). In fact, the matrices

$$\begin{pmatrix} i & 0 \\ 0 & -i \end{pmatrix}, \quad \begin{pmatrix} 0 & -1 \\ 1 & 0 \end{pmatrix}, \quad \begin{pmatrix} 0 & i \\ i & 0 \end{pmatrix},$$

which form a basis of $\mathfrak{su}_2$ over $\mathbf{R}$, also form a basis of $\mathfrak{sl}_2(\mathbf{C})$ over $\mathbf{C}$.

Passing to representations of the Lie algebra $\mathfrak{sl}_2(\mathbf{C})$ is useful because the latter possesses a basis with very convenient commutation relations, namely the one given by the matrices

$$(3) \qquad H = \begin{pmatrix} 1 & 0 \\ 0 & -1 \end{pmatrix}, \quad E_+ = \begin{pmatrix} 0 & 1 \\ 0 & 0 \end{pmatrix}, \quad E_- = \begin{pmatrix} 0 & 0 \\ 1 & 0 \end{pmatrix}$$

with the commutators

$$(4) \qquad [H, E_+] = 2E_+, \quad [H, E_-] = -2E_-, \quad [E_+, E_-] = H.$$

11.3. Theorem 1. *Let $\tau\colon \mathfrak{sl}_2(\mathbf{C}) \to \mathfrak{gl}(V)$ be an irreducible representation of the Lie algebra $\mathfrak{sl}_2(\mathbf{C})$. Then in V there exists a basis $(v_0, v_1, \ldots, v_n)$ such that*

$$\begin{aligned} \tau(H)v_k &= (n-2k)v_k, \\ \tau(E_-)v_k &= v_{k+1}, \qquad (5) \\ \tau(E_+)v_k &= k(n-k+1)v_{k-1}, \end{aligned}$$

where we put $v_{-1} = v_{n+1} = 0$.

Proof. Let v be an arbitrary eigenvector of the operator $\tau(H)$:

$$\tau(H)v = cv.$$

If $\tau(E_+)v \neq 0$, then $\tau(E_+)v$ is again an eigenvector of $\tau(H)$, with corresponding eigenvalue $c+2$:

$$\begin{aligned} \tau(H)\tau(E_+)v &= \tau([H, E_+])v + \tau(E_+)\tau(H)v \\ &= 2\tau(E_+)v + c\tau(E_+)v = (c+2)\tau(E_+)v. \end{aligned}$$

Similarly, one proves that if $\tau(E_-)v \neq 0$, then $\tau(E_-)v$ is an eigenvector of $\tau(H)$ with corresponding eigenvalue $c-2$.

Since $\tau(H)$ can have only finitely many distinct eigenvalues, repeated application of the operator $\tau(E_+)$ to v yields, after a finite number of steps, a vector $v_0 \neq 0$ with the properties

$$\tau(H)v_0 = c_0 v_0, \qquad \tau(E_+)v_0 = 0.$$

Proceeding in a similar fashion with $\tau(E_-)$ and the vector v_0, we finally obtain a vector which is annihilated by $\tau(E_-)$. Now we set

$$v_k = \tau(E_-)^k v_0 \qquad (k = 0, 1, 2, \ldots).$$

Let n be such that $v_0, v_1, \ldots, v_n \neq 0$, whereas $v_{n+1} = 0$.

The vector v_k $(k = 0, 1, \ldots, n)$ is an eigenvector of $\tau(H)$ with corresponding eigenvalue $c_0 - 2k$:

$$\tau(H)v_k = (c_0 - 2k)v_k. \qquad (6)$$

By construction,

$$\tau(E_-)v_k = v_{k+1}. \qquad (7)$$

We use induction on k to show that

$$\tau(E_+)v_k = a_k v_{k-1}, \qquad \text{with } a_k \in \mathbf{C} \qquad (8)$$

(where we put $v_{-1} = 0$). For $k = 0$ this holds by the construction of v_0. For $k = 1$ we have

$$\begin{aligned}\tau(E_+)v_1 &= \tau(E_+)\tau(E_-)v_0 \\ &= \tau([E_+, E_-])v_0 + \tau(E_-)\tau(E_+)v_0 = \tau(H)v_0 = c_0 v_0.\end{aligned}$$

Analogously, letting $k > 1$ and assuming that

$$\tau(E_+)v_{k-1} = a_{k-1}v_{k-2},$$

we obtain

$$\begin{aligned}\tau(E_+)v_k &= \tau(H)v_{k-1} + a_{k-1}\tau(E_-)v_{k-2} \\ &= (c_0 - 2k + 2 + a_{k-1})v_{k-1} = a_k v_{k-1},\end{aligned}$$

where

$$a_k = a_{k-1} + c_0 - 2(k-1). \tag{9}$$

This completes the proof of (8).

The recursion relation (9) together with the condition $a_0 = c_0$ permit us to calculate the coefficient a_k as

$$a_k = kc_0 - 2(1 + 2 + \ldots + (k-1)) = k(c_0 - k + 1). \tag{10}$$

Relations (6)–(8) show that the subspace $\langle v_0, v_1, \ldots, v_n\rangle$ is invariant under the representation τ. Since τ is irreducible, we conclude that

$$\langle v_0, v_1, \ldots, v_n\rangle = V.$$

To complete the proof of the theorem it remains to show that $c_0 = n$. The simplest way to do this is to use the fact that relations (9) and (10) also hold for $k = n + 1$. But since $v_{n+1} = 0$, we get

$$a_{n+1} = (n+1)(c_0 - n) = 0,$$

whence $c_0 = n$. □

Corollary. *Two irreducible representations of $\mathfrak{sl}_2(\mathbf{C})$ having the same dimension are isomorphic.*

Proof. . In fact, if the dimension $n + 1$ of representation τ is known, then τ is completely described by formulas (5). □

11.4. Theorem 2. *Every irreducible representation T of the Lie group* SU_2 *is isomorphic to one of the representations Φ_n constructed in* 7.4.

PROOF. By the Corollary to Theorem 1 of 10.4, it suffices to show that the tangent representation d_eT of the Lie algebra $\mathfrak{su}_2$ is isomorphic to one of the representations $d_e\Phi_n$. According to the Corollary to Theorem 2 of 10.6, both d_eT and $d_e\Phi_n$ are irreducible. We extend them to the complexification $\mathfrak{sl}_2(\mathbf{C})$ of $\mathfrak{su}_2$ and then apply the Corollary to Theorem 1 of 11.3. For $n+1 = \dim T$ we conclude that $d_eT \simeq d_e\Phi_n$. □

For another proof of Theorem 2, based on the theory of characters, see Exercise 7 to Section 8.

Corollary 1. *Every irreducible representation T of the Lie group* SO_3 *is isomorphic to one of the representations Ψ_n constructed in* 7.4.

PROOF. In fact, T is obtained through factorization from one of the irreducible representations of SU_2. By Theorem 2, the latter must be isomorphic to one of the representations Φ_n, and then $T \simeq \Psi_n$. □

Corollary 2. *The representation of* SO_3 *in the space U_m of functions on the sphere* (see Section 9) *is isomorphic to Ψ_{2m}.* □

11.5. In Section 9 we gave without proof formulas expressing the *Laplace spherical functions* $Y_{m,k}$, $k \neq 0$, through Legendre polynomials. We now prove these formulas using Theorem 1 of this section.

We shall regard the three-dimensional Euclidean space in which the sphere S lies as the space $\mathbf{E}$ of traceless Hermitian matrices of order two (see 7.2), and accordingly regard SO_3 as the group of rotations in $\mathbf{E}$.

Next, we fix in $\mathbf{E}$ the orthonormal basis (E_1, E_2, E_3), where

$$E_1 = \begin{pmatrix} 0 & 1 \\ 1 & 0 \end{pmatrix}, \qquad E_2 = \begin{pmatrix} 0 & i \\ -i & 0 \end{pmatrix}, \qquad E_3 = \begin{pmatrix} -1 & 0 \\ 0 & 1 \end{pmatrix}.$$

Our calculations will be carried out in the coordinates x_1, x_2, x_3 relative to this basis. Then the "north pole" of the sphere S is the matrix E_3.

Given m, we consider the irreducible representation of SO_3 in the space U_m of functions on S.

We lift it to SU_2 by means of the homomorphism $P\colon SU_2 \to SO_3$ constructed in 7.4, and we denote the resulting representation of SU_2 by T. The tangent representation $d_e T$ of $\mathfrak{su}_2$ and its extension to $\mathfrak{sl}_2(\mathbf{C})$ will both be denoted by τ.

By Theorem 1, the representation τ is given by formulas (5) relative to some basis $(v_0, v_1, \ldots, v_n)$ of U_m, where $n = 2m$ (since $\dim U_m = 2m+1$). Our first objective is to show that $(v_0, v_1, \ldots, v_n)$ coincides with the basis $(Y_{m,m}, Y_{m,m-1}, \ldots, Y_{m,-m})$ up to a normalization. To this end we must show that the spherical functions $Y_{m,k}$ are eigenfunctions of the operator $\tau(H)$.

Lemma. $\tau(H)Y_{m,k} = 2kY_{m,k}$.

PROOF. As in Section 7, we put

$$A(z) = \begin{pmatrix} z & 0 \\ 0 & z^{-1} \end{pmatrix} \in \mathrm{SU}_2 \qquad (z \in \mathbf{C},\ |z| = 1).$$

The matrix $iH \in \mathfrak{su}_2$ is the tangent vector to the curve $g(t) = A(\mathrm{e}^{it})$ at $t = 0$. By the definition of the tangent representation,

$$\tau(iH)Y_{m,k} = \frac{d}{dt}T(g(t))Y_{m,k}\big|_{t=0}.$$

In the course of the proof of the Theorem of 7.2 it was established that $P(A(\mathrm{e}^{it}))$ is the rotation through an angle of $2t$ around the axis $\langle H \rangle = \langle E_3 \rangle$:

$$P(A(\mathrm{e}^{it})) = h(\mathrm{e}^{2it})$$

(for the notation h, see 9.4). Therefore,

$$T(g(t))Y_{m,k} = h(\mathrm{e}^{2it})_* Y_{m,k} = \mathrm{e}^{2kit}Y_{m,k}$$

(see Theorem 2, 9.5), and

$$\tau(iH)Y_{m,k} = \frac{d}{dt}(\mathrm{e}^{2kit}Y_{m,k})\big|_{t=0} = 2kiY_{m,k},$$

from which the needed relation follows upon dividing both sides by i. □

Since v_k is an eigenvector of $\tau(H)$ corresponding to the eigenvalue $n - 2k = 2(m-k)$, it is proportional to the spherical function $Y_{m,m-k}$. It then follows from (5) that for a suitable normalization of the spherical functions we have

$$Y_{m,k} = \tau(E_+)^k Y_{m,0} \qquad (k > 0) \tag{11}$$

and

$$Y_{m,-k} = \tau(E_-)^k Y_{m,0} \qquad (k > 0). \tag{12}$$

11.6. Now let us find the *explicit form of the operators* $\tau(E_+)$ and $\tau(E_-)$ in U_m. To this end we apply the Proposition of 10.11 to the action of SU_2 on the sphere S obtained by composing the usual action of SO_3 on S and the homomorphism P. This allows us to represent the operators $\tau(C)$, with $C \in \mathfrak{su}_2$, as differential operators on U_m. Specifically,

$$\tau(C)f = -\partial_{ds(C)}f \qquad (f \in U_m),$$

where $ds(C)$ is the velocity field of the action s corresponding to C.

Since s is the restriction of the linear representation P to the sphere S,

$$ds(C)X = dP(C)X = CX - XC$$

for all $X \in S \subset \mathbf{E}$. (Here we think of the tangent space to S at the point X as being canonically embedded as a subspace in $\mathbf{E}$.) Thus, for arbitrary $C \in \mathfrak{su}_2$, $X \in S$, and $f \in U_m$ we have

$$(\tau(C)f)(X) = (\partial_{XC-CX}f)(X). \tag{13}$$

This formula extends by linearity to arbitrary matrices $C \in \mathfrak{sl}_2(\mathbf{C})$. In particular, it can be applied to the matrices E_+ and E_- of interest to us.

Let

$$X = x_1E_1 + x_2E_2 + x_3E_3 = \begin{pmatrix} -x_3 & x_1 + ix_2 \\ x_1 - ix_2 & x_3 \end{pmatrix} \in \mathbf{E}.$$

A straightforward calculation shows that

$$\begin{aligned} XE_+ - E_+X &= \begin{pmatrix} -x_1 + ix_2 & -2x_3 \\ 0 & x_1 - ix_2 \end{pmatrix} \\ &= (x_1 - ix_2)E_3 - x_3(E_1 - iE_2). \end{aligned}$$

Hence, if $f \in U_m$ is the restriction of a function F on $\mathbf{E}$ to the sphere S, then $\tau(E_+)f$ is the restriction to S of the function

$$DF = (x_1 - ix_2)\frac{\partial F}{\partial x_3} - x_3\Big(\frac{\partial F}{\partial x_1} - i\frac{\partial F}{\partial x_2}\Big).$$

The spherical function $Y_{m,0}$ is the restriction to S of $F = P_m(x_3)$, where P_m is the Legendre polynomial of degree m. We have

$$DF = (x_1 - ix_2)P_m'(x_3).$$

Applying the operator D once more, the second term vanishes, because

$$\Big(\frac{\partial}{\partial x_1} - i\frac{\partial}{\partial x_2}\Big)(x_1 - ix_2) = 0.$$

The same holds true for the succeeding iterations. Therefore,

$$D^k F = (x_1 - ix_2)^k P_m^{(k)}(x_3),$$

and so

$$Y_{m,k} = (\xi_1 - i\xi_2)^k P_m^{(k)}(\xi_3) \qquad (k > 0).$$

The second of formulas (3) of 9.6 is established in an analogous manner.

Appendices

A1. Presentation of Groups By Means of Generators and Relations

A1.1. Let S be a subset of the group G. We call any finite seqence $w = (x_1^{\varepsilon_1}, \dots, x_m^{\varepsilon_m})$, where $x_i \in S$ and $\varepsilon_i = \pm 1$, an S-WORD, including in this definition the "empty" word $\emptyset$, for which $m = 0$. Here the exponents ε_i should be interpreted formally: in other words, if $x \in S$ and $x^{-1} = y \in S$, then the "letters" x^{-1} and y should nevertheless be regarded as distinct. We let $W(S)$ denote the set of all S-words. The PRODUCT of two words is formed by writing them successively:

$$(x_1^{\varepsilon_1}, \dots, x_m^{\varepsilon_m})(y_1^{\eta_1}, \dots, y_n^{\eta_n}) = (x_1^{\varepsilon_1}, \dots, x_m^{\varepsilon_m}, y_1^{\eta_1}, \dots, y_n^{\eta_n}).$$

It is obvious that this operation is associative and admits the empty word $\emptyset$ as a neutral element.

By assigning to each word the product of its "letters" we obtain the mapping

$$(1) \qquad p: W(S) \to G, \qquad (x_1^{\varepsilon_1}, \dots, x_m^{\varepsilon_m}) \mapsto x_1^{\varepsilon_1} \dots x_m^{\varepsilon_m}.$$

We take $p(\emptyset)$ to be the identity e of G. Obviously,

$$(2) \qquad p(w_1 w_2) = p(w_1)p(w_2).$$

For $w = (x_1^{\varepsilon_1}, \dots, x_m^{\varepsilon_m})$ we put $\bar{w} = (x_m^{-\varepsilon_m}, \dots, x_1^{-\varepsilon_1})$. Then

$$(3) \qquad p(\bar{w}) = p(w)^{-1}.$$

It follows from (2) and (3) that $p(W(S))$ is a subgroup of G. It is called the SUBGROUP GENERATED BY THE SET S, and is denoted here by $\langle S \rangle$. In particular, *if S reduces to a single element x, then $\langle S \rangle = \langle x \rangle$ is the cyclic subgroup generated by the element x.*

When $\langle S \rangle = G$ we say that S is a SYSTEM OF GENERATORS of the group G (or that S GENERATES G). Every group possesses various systems of generators, among them the rather uninteresting one $S = G$.

E.B. Vinberg, *Linear Representations of Groups*, Modern Birkhäuser Classics,
DOI 10.1007/978-3-0348-0063-1_6, © Birkhäuser Verlag 1989

What is of practical value is that a group containing a large number, or even infinitely many elements, may nevertheless admit a system of generators consisting of a small number of elements. For instance, for any order n, the cyclic group C_n is by definition generated by a single element. The dihedral group D_n, isomorphic to the symmetry group of a regular polygon with n sides, is generated by two elements x and y, representing respectively a rotation through an angle of $2\pi/n$ and a reflection in the line passing through the origin of the polygon and one of its vertices.

A1.2. Let S be a fixed system of generators of the group G. Every element $g \in G$ can be written in the form $g = x_1^{\varepsilon_1} \dots x_m^{\varepsilon_m}$, with $x_i \in S$ and $\varepsilon_i = \pm 1$. Such a representation, however, is not unique (except for the case $S = \emptyset$). In other words, the map $p\colon W(S) \to G$ constructed in A1.1 is not injective.

We define an *equivalence relation on* $W(S)$ setting

$$(4) \qquad w_1 \sim w_2 \qquad \text{if } p(w_1) = p(w_2).$$

Each true assertion "$w_1 \sim w_2$" is called a RELATION AMONG THE GENERATORS in the system S.

Examples.

1. There are always the trivial relations $(x, x^{-1}) \sim \emptyset$ and $(x^{-1}, x) \sim \emptyset$.

2. The relation

$$\underbrace{(x, \dots, x)}_{n} \sim \emptyset$$

holds in the cyclic group $C_n = \langle x \rangle$.

3. The following relations hold in the dihedral group $D_n = \langle x, y \rangle$ (see A1.1):

$$\underbrace{(x, \dots, x)}_{n} \sim \emptyset, \qquad (y, y) \sim \emptyset, \qquad (x, y) \sim (y, x^{-1}).$$

If all relations are known, then one can reconstruct the group G from the given system of generators S. Specifically, the elements of G can be identified with the equivalence classes of relation (4) in $W(S)$, and then the product of any two elements can be found by multiplying words, i.e., representatives of the equivalence classes.

A1.3. Actually, it is not necessary to indicate all relations explicitly: they can be derived from only part of them. The derivation uses the symmetry and transitivity of the equivalence relation as well as the following property, which is a consequence of (2):

($*$) if $w_1 \sim w_2$ and u, v are arbitrary words, then

$$uw_1v \sim uw_2v.$$

More precisely, let R be a set of relations. An R-ELEMENTARY TRANSFORMATION is defined as the transition from the word uw_1v to the equivalent word uw_2v, where $u, v, w_1, w_2 \in W(S)$, $w_1 \sim w_2$, and one of the relations $w_1 \sim w_2$, $w_2 \sim w_1$ either is a trivial relation of the type $(x, x^{-1}) \sim \emptyset$ or $(x^{-1}, x) \sim \emptyset$, or else belongs to R. Depending on which of the two cases occurs, one speaks of an elementary transformation of the FIRST or of the SECOND TYPE. Two words w_1 and w_2 are said to be R-EQUIVALENT if w_2 can be obtained from w_1 through a chain of R-elementary transformations.

The set R is called a SYSTEM OF DEFINING RELATIONS if any two equivalent words are R-equivalent. This is the precise meaning of the assertion that every relation can be derived from the relations of the system R.

Examples.

1. We prove that

$$(\underbrace{x, \dots, x}_{n}) \sim \emptyset$$

is a defining relation for the cyclic group $C_n = \langle x \rangle$, or, more precisely, that the set R consisting of only the indicated relation is a system of defining relations for C_n.

By means of R-elementary transformations of the first type (i.e., "cancellations" of adjacent letters x and x^{-1}) one can bring every word $w \in W(\{x\})$ to one of the forms $(x, \dots, x)$, $(x^{-1}, \dots, x^{-1})$, or $\emptyset$. Next, by means of R-elementary operations of the second kind one can, by erasing the word $(\underbrace{x, \dots, x}_{n})$ or writing it and subsequently doing some cancellations, pass to a word of the form $(\underbrace{x, \dots, x}_{k})$ with $0 \leq k < n$.

Suppose now that $w_1 \sim w_2$, where w_1 and w_2 are R-equivalent to the words $(\underbrace{x, \dots, x}_{k})$, $0 \leq k < n$, and $(\underbrace{x, \dots, x}_{\ell})$, $0 \leq \ell < n$, respectively. Then $p(w_1) =$

x^k, $p(w_2) = x^\ell$, and since $p(w_1) = p(w_2)$, we get that $k = \ell$. Hence, the words w_1 and w_2 are R-equivalent.

2. In an analogous manner one proves that the set R of relations among the generators x, y of the dihedral group D_n indicated in Example 3 of A1.2 is a system of defining relations for D_n. To this end one verifies first that each word $w \in W(\{x,y\})$ is R-equivalent to a word of the type

$$\underbrace{(x,\dots,x)}_{k} \quad \text{or} \quad \underbrace{(x,\dots,x}_{k},y),$$

where $0 \le k < n$. Then one uses the fact that the elements of D_n can be written uniquely in the form x^k or $x^k y$, with $0 \le k < n$.

In practice one usually adopts a simplified convention for writing words and relations. Specifically, the word $w = (x_1^{\varepsilon_1}, \dots, x_m^{\varepsilon_m})$ is written as the product $x_1^{\varepsilon_1} \dots x_m^{\varepsilon_m}$, which is understood not only as the element of the group equal to it, but also as the way in which that element is obtained from generators. Accordingly, relations among generators are written as equalities in the group. A group with given generators and defining relations is denoted with the symbol $\langle \dots \mid \dots \rangle$, where one writes the generators to the left of the vertical bar and the defining relations to the right. For example,

$$C_n = \langle x \mid x^n = e \rangle$$

and

$$D_n = \langle x, y \mid x^n = e,\ y^2 = e,\ xy = yx^{-1} \rangle.$$

We give without proof two more complicated examples.

3. $S_n = \langle \tau_1, \dots, \tau_{n-1} \mid \tau_i^2 = \varepsilon,\ \tau_i\tau_j = \tau_j\tau_i \text{ for } |i-j| > 1,\ \tau_{i+1}\tau_i\tau_{i+1} = \tau_i\tau_{i+1}\tau_i \rangle$, where $\tau_i = (i, i+1)$.

4. $\mathrm{SL}_2(\mathbf{Z}) = \langle A, B \mid A^4 = E,\ B^3 = E,\ A^2B = BA^2 \rangle$, where

$$A = \begin{pmatrix} 0 & -1 \\ 1 & 0 \end{pmatrix}, \qquad B = \begin{pmatrix} 0 & 1 \\ -1 & -1 \end{pmatrix}.$$

A1.4. Generators and defining relations can be used not only to describe groups already given by some other means, but also to define new groups.

Let S be an arbitrary set. We call any finite sequence of symbols of the form x or x^{-1}, where $x \in S$, an S-WORD. Next, we define the PRODUCT OF TWO S-WORDS as the word obtained by writing them successively. In this manner we obtain a monoid $W(S)$ (i.e., a set endowed with an associative operation) in which the empty word serves as a neutral element.

Further, let R be an arbitarary set of "relations" of the type $w_1 \sim w_2$, where $w_1, w_2 \in W(S)$. We define the R-ELEMENTARY TRANSFORMATIONS and R-EQUIVALENCE of words as in A1.3. The class of R-equivalent words containing w will be denoted by $p(w)$.

Now we define the MULTIPLICATION OF CLASSES by the rule

$$p(w_1)p(w_2) = p(w_1 w_2).$$

It is readily seen that this definition is correct, i.e., if $w_1 \sim w_1'$ and $w_2 \sim w_2'$, then $w_1 w_2 \sim w_1' w_2'$. That class multiplication is associative follows from the associativity of word multiplication. The class $p(\emptyset)$ is a neutral element for class multiplication. Finally, if one defines the word $\bar{w}$ as in A1.1 above, then $w\bar{w}$ and $\bar{w}w$ can be reduced to the empty word through elementary transformations of the first type, and so the class $p(\bar{w})$ is the inverse of $p(w)$.

Thus, *the equivalence classes of words form a group* that we will denote by G.

Since every word can be expressed as a product of one-letter words (x) and (x^{-1}), G is generated by the elements $p((x))$, $x \in S$. Two products of the type

$$p((x_1))^{\varepsilon_1} \dots p((x_m))^{\varepsilon_m} \qquad (\text{with } \varepsilon_i = \pm 1)$$

are equal if and only if the corresponding words

$$(x_1^{\varepsilon_1}, \dots, x_m^{\varepsilon_m})$$

are equivalent. In particular, to every "elementary" equivalence relation belonging to the set R there corresponds a relation among the generators $p((x))$ of the group G; all other relations are, according to the definition, consequences of these.

Unfortunately, the question of whether some arbitrarily given relation holds in G, i.e., of whether or not it can be derived from the relations in R, cannot, in general, be answered. In the general case one cannot even decide whether the group G contains more than one element. In those particular cases where these questions can be settled, an individual approach is required.

A1.5. If generators and defining relations for a group G are known, one can describe all group homomorphisms of G into an arbitrary group H.

Specifically, let S be a system of generators of G. For any homomorphism $\phi: G \to H$,

$$(5) \qquad \phi(x_1^{\varepsilon_1} \dots x_m^{\varepsilon_m}) = \phi(x_1)^{\varepsilon_1} \dots \phi(x_m)^{\varepsilon_m}.$$

Since every element of G is expressible as $x_1^{\varepsilon_1} \dots x_m^{\varepsilon_m}$, with $x_i \in S$, formula (5) shows that *the homomorphism ϕ is uniquely determined by its restriction to S.*

Next, let R be a system of defining relations for G. *A necessary and sufficient condition for a map $\psi: S \to H$ to admit an extension to a group homomorphism $\phi: G \to H$ is that*

$$\psi(x_1)^{\varepsilon_1} \dots \psi(x_m)^{\varepsilon_m} = \psi(y_1)^{\eta_1} \dots \psi(y_n)^{\eta_n}$$

for every relation

$$x_1^{\varepsilon_1} \dots x_m^{\varepsilon_m} = y_1^{\eta_1} \dots y_n^{\eta_n}$$

belonging to R.

The necessity of this condition is obvious. Let us prove its sufficiency.

Suppose we have a map $\psi: S \to H$ which satisfies the indicated condition. Consider the monoid $W(S)$, and define a homomorphism q of $W(S)$ into H by the rule

$$q(x_1^{\varepsilon_1} \dots x_m^{\varepsilon_m})) = \psi(x_1)^{\varepsilon_1} \dots \psi(x_m)^{\varepsilon_m}.$$

If the system R contains a relation $w_1 \sim w_2$, then by the above condition $q(w_1) = q(w_2)$ and hence $q(uw_1v) = q(uw_2v)$ for arbitrary words u, v. The same holds true if $w_1 \sim w_2$ is a trivial relation. Thus, words that can be obtained from one another through R-elementary transformations have the same image under q. Since R is a system of defining relations, it follows that equivalent words have the same image under q. Consequently, we can write $q = \phi \circ p$, where ϕ is a map of G into H. It is readily verified that ϕ is a homomorphism that extends the map ψ.

Examples.

1. Any homomorphism ϕ of the cyclic group $C_n = \langle x \rangle$ into an arbitrary group H is uniquely specified by the single element $\phi(x) = h$, and the latter can be any element of H satisfying the requirement $h^n = e$.

2. Let us find *all complex irreducible representations of the dihedral group*

$$D_n = \langle x, y \mid x^n = y^2 = e,\ xy = yx^{-1} \rangle.$$

A linear representation of D_n is a homomorphism $T: D_n \to \mathrm{GL}(V)$. It is uniquely determined by specifying the linear operators $T(x)$ and $T(y)$ subject to the conditions

$$\begin{cases} T(x)^n = T(y)^2 = \varepsilon, \\ T(x)T(y) = T(y)T(x)^{-1}. \end{cases} \tag{6}$$

Let v be an eigenvector of $T(x)$:

$$T(x)v = cv, \qquad \text{with } c \in \mathbf{C}.$$

Then $T(y)v$ is again an eigenvector of $T(x)$:

$$T(x)T(y)v = T(y)T(x)^{-1}v = c^{-1}T(y)v.$$

Since $T(y)^2v = v$, the subspace $\langle v, T(y)v\rangle$ is T-invariant.

Suppose now that T is an irreducible representation. Then

$$V = \langle v, T(y)v\rangle.$$

If $c \neq \pm 1$, then $c^{-1} \neq c$, and the vectors v and $T(y)v$ are linearly independent. In this case $\dim V = 2$, and in the basis $(v, T(y)v)$ the operators $T(x)$ and $T(y)$ are given by the matrices

$$\begin{pmatrix} c & 0 \\ 0 & c^{-1} \end{pmatrix} \qquad \text{and} \qquad \begin{pmatrix} 0 & 1 \\ 1 & 0 \end{pmatrix},$$

respectively. Relations (6) are satisfied provided $c^n = 1$. We thus obtain $\left[\frac{1}{2}(n-1)\right]$ two-dimensional irreducible representations of the group D_n.

If now $c = \pm 1$, then $c^{-1} = c$ and $T(x)$ is a scalar operator. Let w be an eigenvector of $T(y)$:

$$T(y)w = \pm w.$$

Then $\langle w\rangle$ is an invariant subspace, and so $V = \langle v\rangle$. We thus obtain four one-dimensional representations of D_n for n even and two for n odd.

A1.6. Let us examine in more detail the *abelian groups with finitely many generators.*

Let G be an abelian group with a system of generators $S = \{x_1, \dots, x_n\}$. We shall assume that the system of defining relations of G includes the commutativity relations

$$x_i x_j = x_j x_i.$$

Using these, every product of generators can be reduced to the form

$$x_1^{k_1} \dots x_n^{k_n} \qquad (\text{with } k_1, \dots, k_n \in \mathbf{Z}).$$

Hence, we may assume that all the other relations have the form

$$x_1^{k_{s1}} \dots x_n^{k_{sn}} = e.$$

The matrix $K = [k_{ij}]$ with entries $k_{ij} \in \mathbf{Z}$ specifies G completely.

The following "elementary" transformations of the system of defining relations for G (except for the commutativity relations) are admissible:

1) multiplication of a relation by another one (with subsequent ordering of the factors);

2) replacement of the relation $w = e$ by $w^{-1} = e$;

3) permutation of relations.

To 1)–3) there correspond *the elementary row operations on K over the integers.* To analogous elementary operations on the system of generators of G there correspond *the elementary column operations on K over the integers.*

It is known that every matrix with entries in $\mathbf{Z}$ can be reduced by elementary row and column operations over $\mathbf{Z}$ to diagonal form. Consequently, in every group G with a finite number of generators there exists a system of generators $S = \{x_1, \dots, x_n\}$ such that for the system of defining relations that $x_1, \dots, x_n$ satisfy one can choose

$$\begin{cases} x_i x_j = x_j x_i, \\ x_i^{k_i} = e, \qquad \text{with } k_i \in \mathbf{Z},\ i = 1, \dots, m. \end{cases}$$

A group with such defining relations is the direct product of the cyclic groups $\langle x_i \rangle$, $i = 1, \dots, n$. We have thus established the following result.

Theorem. *Every abelian group with a finite number of generators decomposes into a direct product of cyclic groups.*

A2. Tensor Products

A2.1. Let V and U be vector spaces over the same field F. Consider the additive abelian group A generated by the symbols $x \otimes y$ $(x \in V, y \in U)$ with the following defining relations (which, of course, supplement the commutativity relations):

(1) $$(x_1 + x_2) \otimes y = x_1 \otimes y + x_2 \otimes y,$$
(2) $$x \otimes (y_1 + y_2) = x \otimes y_1 + x \otimes y_2,$$
(3) $$cx \otimes y = x \otimes cy \qquad (c \in F).$$

From (1)–(3) it follows, in particular, that $0 \otimes 0$ is the null (neutral) element of A, and that $(-x) \otimes y$ is the opposite of $x \otimes y$. Every element of A can be written as a sum

$$\sum x_i \otimes y_i, \qquad \text{with } x_i \in V,\ y_i \in U.$$

Basic Lemma. *Let Γ be a map of the direct product $V \times U$ into an additive abelian group W, with the properties*

$$\begin{array}{ll} 1) & \Gamma(x_1 + x_2, y) = \Gamma(x_1, y) + \Gamma(x_2, y); \\ 2) & \Gamma(x, y_1 + y_2) = \Gamma(x, y_1) + \Gamma(x, y_2); \\ 3) & \Gamma(cx, y) = \Gamma(x, cy) \qquad (c \in F). \end{array}$$

Then there exists a homomorphism $\gamma\colon A \to W$ such that

$$\gamma\left(\sum x_i \otimes y_i\right) = \sum \Gamma(x_i, y_i) \tag{4}$$

for all $x_i \in V$, $y_i \in U$.

PROOF. It follows from the properties 1)–3) of the map Γ that the right-hand side of (4) is not affected by elementary transformations of the sum $\sum x_i \otimes y_i$ (see A1) which use the defining relations (1)–(3) of the group A. Hence, formula (4) can be taken as the definition of the map $\gamma\colon A \to W$. That γ is a homomorphism is plain. □

Now define the *multiplication of the elements of the group A by scalars in F* by the rule

$$a \sum x_i \otimes y_i = \sum a x_i \otimes y_i \qquad (a \in F). \tag{5}$$

The correctness of this definition follows from the Basic Lemma, applied to the map $\Gamma\colon V \times U \to A$ given by

$$\Gamma(x, y) = ax \otimes y.$$

That conditions 1)–3) of the lemma are satisfied is guaranteed by the following equalities, which in turn are consequences of the definition of A:

$$\begin{aligned} a(x_1 + x_2) \otimes y &= ax_1 \otimes y + ax_2 \otimes y, \\ ax \otimes (y_1 + y_2) &= ax \otimes y_1 + ax \otimes y_2, \end{aligned}$$

and

$$a(cx) \otimes y = ax \otimes cy.$$

It is readily verified that the operation of multiplication by elements of F defined in the indicated manner turns A into a *vector space* over F. With this structure A is called the TENSOR PRODUCT OF THE VECTOR SPACES V AND U, and is denoted by $V \otimes U$.

From the Basic Lemma one derives

Theorem 1. *Let Γ be a bilinear map of the product $V \times U$ into a vector space W over F. Then there exists a linear map $\gamma: V \otimes U \to W$ such that*

$$\gamma(\sum x_i \otimes y_i) = \sum \Gamma(x_i, y_i) \tag{6}$$

for all $x_i \in V$, $y_i \in U$.

PROOF. In fact, a bilinear map trivially satisfies conditions 1)–3) of the Basic Lemma, and the map γ that the latter provides is here linear and not merely a group homomorphism. □

It is precisely Theorem 1 that we are going to use in the ensuing discussion.

A2.2. Theorem 2. *Let $(e) = (e_1, \dots, e_n)$ and $(f) = (f_1, \dots, f_n)$ be bases of the spaces V and U respectively. Then the vectors $e_i \otimes f_j$ ($i = 1, \dots, n$; $j = 1, \dots, m$) constitute a basis of $V \otimes U$.*

PROOF. Let $x = \sum x_i e_i \in V$ and $y = \sum b_j f_j \in U$. Then

$$x \otimes y = \sum_{i,j} a_i b_j (e_i \otimes f_j). \tag{7}$$

We see that the vectors of the form $x \otimes y$, and consequently all vectors in $V \otimes U$ can be written as linear combinations of $e_i \otimes f_j$.

To show that the nm vectors $e_i \otimes f_j$ are linearly independent, we first give a definition. Let ε and η be linear functions on V and on U respectively. Then the linear function $\varepsilon \otimes \eta$ on $V \otimes U$ is defined by the formula

$$(\varepsilon \otimes \eta)(\sum x_i \otimes y_i) = \sum \varepsilon(x_i)\eta(y_i). \tag{8}$$

The correctness of this definition is guaranteed by Theorem 1, applied to the bilinear function $\Gamma(x, y) = \varepsilon(x)\eta(y)$.

Suppose now that there is a linear dependence relation among the vectors $e_i \otimes f_j$:

$$\sum c_{ij}(e_i \otimes f_j) = 0.$$

Let $(\varepsilon_1, \dots, \varepsilon_n)$ and $(\eta_1, \dots, \eta_n)$ be bases of the spaces V' and U' dual to the bases (e) and (f) of V and U respectively. Then

$$(\varepsilon_k \otimes \eta_\ell)(e_i \otimes f_j) = \varepsilon_k(e_i)\eta_\ell(f_j) = \begin{cases} 1, & \text{if } i = k \text{ and } j = \ell, \\ 0, & \text{otherwise,} \end{cases}$$

and so

$$(\varepsilon_k \otimes \eta_\ell)\Big(\sum_{i,j} c_{ij}(e_i \otimes f_j)\Big) = c_{kl} = 0$$

for all k and ℓ, as we needed to show. □

Corollary. $\dim V \otimes U = \dim V \cdot \dim U$. □

Theorem 2 also yields the following two decompositions of the tensor product $V \otimes U$ into a direct sum of subspaces:

$$V \otimes U = (e_1 \otimes U) \oplus \ldots \oplus (e_n \otimes U)$$

and

$$V \otimes U = (V \otimes f_1) \oplus \ldots \oplus (V \otimes f_m).$$

The summands of the first (second) decomposition are the linear spans of the rows (respectively, columns) of the matrix with vector entries $(e_i \otimes f_j)$.

A2.3. In the notations of A2.2, every element $z \in V \otimes U$ is uniquely expressible as

(9) $$z = \sum_{i,j} c_{ij}(e_i \otimes f_j).$$

We shall refer to $C = [c_{ij}]$ as the MATRIX OF z relative to the bases (e) and (f).

The elements of the form $x \otimes y$ with $x \in U$ and $y \in V$ are called the DECOMPOSABLE ELEMENTS of the tensor product $V \otimes U$. Formula (7) shows that the matrix of a decomposable element can be written as $[a_i b_j]$ for certain $a_i, b_j \in F$, and consequently has rank ≤ 1. Conversely, every matrix of rank ≤ 1 can be written in the indicated form and hence is the matrix of a decomposable element.

Examining the matrix C of a nonnull decomposable element, it is readily established that the representation $x \otimes y$ of the latter is unique up to simultaneous multiplication of x and y by c and c^{-1} respectively, where $c \neq 0$ is an arbitrary scalar. In fact, the coordinates of x (respectively, y) must be proportional to the entries of any nonnull column (respectively, row) of C.

A2.4. We wish to draw attention to the case where one of the spaces V, U is *one-dimensional.* Suppose, for example, that $\dim U = 1$, and $f \neq 0$ is an arbitrary vector in U. Then

$$V \otimes U = V \otimes f,$$

and the map $x \mapsto x \otimes f$ $(x \in V)$ is a linear isomorphism of V onto $V \otimes U$.

A2.5. With each pair of linear operators $\alpha \in \mathrm{L}(V)$ and $\beta \in \mathrm{L}(U)$ one can associate the linear operator $\alpha \otimes \beta \in \mathrm{L}(V \otimes U)$ acting according to the rule

$$(10) \qquad (\alpha \otimes \beta)(\sum x_i \otimes y_i) = \sum \alpha(x_i) \otimes \beta(y_i).$$

The correctness of this definition is again guaranteed by Theorem 1, applied to the bilinear map $\Gamma(x, y) = \alpha(x) \otimes \beta(y)$. Let $A = [a_{ik}]$ and $B = [b_{j\ell}]$ be the matrices of the operators α and β in the bases (e) and (f) respectively. Then

$$(\alpha \otimes \beta)(e_k \otimes f_\ell) = \alpha(e_k) \otimes \beta(f_\ell) = \sum_{i,j} a_{ik} b_{j\ell} (e_i \otimes f_j),$$

so that the matrix elements of the operator $\alpha \otimes \beta$ are the various possible products of matrix elements of the operators α and β. Upon ordering the basis vectors $e_i \otimes f_j$ first with respect to the index i and then with respect to the index j, the matrix of $\alpha \otimes \beta$ in the resulting ordered basis takes on the block form

$$\begin{pmatrix} a_{11}B & a_{12}B & \cdots & a_{1n}B \\ a_{21}B & a_{22}B & \cdots & a_{2n}B \\ \cdots & \cdots & \cdots & \cdots \\ a_{n1}B & a_{n2}B & \cdots & a_{nn}B \end{pmatrix}.$$

From this it follows, in particular, that

$$\operatorname{tr}(\alpha \otimes \beta) = \operatorname{tr} \alpha \cdot \operatorname{tr} \beta.$$

A2.6. *The space $V \otimes V'$ is canonically isomorphic to* $\mathrm{L}(V)$. Specifically, to each element $z = \sum v_i \otimes f_i \in V \otimes V'$ ($v_i \in V$, $f_i \in V'$) we assign the linear operator

$$\alpha_z \colon x \mapsto \sum f_i(x) v_i$$

acting in V. The correctness of this definition is guaranteed by Theorem 1, applied, for each fixed x, to the bilinear map

$$\Gamma(v, f) = f(x)v \qquad (v \in V, f \in V').$$

The linearity of α_z is obvious.

Let $(e) = (e_1, \ldots, e_n)$ be a basis the V, and $(\varepsilon) = (\varepsilon_1, \ldots, \varepsilon_n)$ the dual basis of V'. According to our definition, to

$$z = \sum_{i,j} c_{ij}(e_i \otimes \varepsilon_j) = \sum_i \Big(\sum_j c_{ij} e_i \otimes \varepsilon_j\Big) \in V \otimes V'$$

there corresponds the linear operator

$$\alpha_z \colon x \mapsto \sum_{i,j} c_{ij} \varepsilon_j(x) e_i.$$

In particular,

$$\alpha_z(e_j) = \sum_i c_{ij} e_i,$$

and so the matrix of α_z in the basis (e) coincides with $[c_{ij}]$. This shows, in particular, that the map $z \mapsto \alpha_z$ is a linear isomorphism of $V \otimes V'$ onto $\mathrm{L}(V)$. With this isomorphism in mind, the spaces $V \otimes V'$ and $\mathrm{L}(V)$ are usually identified.

By the above definition, to each decomposable element $v \otimes f \in V \otimes V'$ there corresponds the linear operator of rank ≤ 1 acting as

$$x \mapsto f(x)v.$$

Its trace equals $f(v)$.

A2.7. *The space $V' \otimes V'$ is canonically isomorphic to the space $\mathrm{B}(V)$ of bilinear functions on V.* Specifically, with each element $z = \sum_i f_i \otimes g_i \in V' \otimes V'$ ($f_i, g_i \in V'$) we associate the bilinear function

$$F_z(x, y) = \sum f_i(x) g_i(y).$$

The correctness of this definition is verified with the help of Theorem 1. The matrix of z in the basis of $V' \otimes V'$ provided by the vectors $\varepsilon_i \otimes \varepsilon_j$ coincides with the matrix of the bilinear function F_z in the basis (e) of V.

The various tensor products of the form

$$\underbrace{V \otimes \ldots \otimes V}_{k} \otimes \underbrace{V' \otimes \ldots \otimes V'}_{\ell}$$

admit analogous interpretations. For instance, the elements of $V \otimes V' \otimes V'$ may be viewed as bilinear operations in the space V.

A3. The Convex Hull of a Compact Set

Here we prove the following theorem that was used in Section 2: *the convex hull of any compact set $K \subset \mathbf{R}^n$ is compact (and hence closed).*

We recall the definition: the convex hull of the set $K \subset \mathbf{R}^n$ is the set

$$\operatorname{conv} K = \Big\{ \sum_{i=1}^{m} c_i x_i \mid x_i \in K,\ c_i \geq 0,\ \sum_{i=1}^{m} c_i = 1 \Big\}.$$

Here m is an arbitrary positive integer, i.e., sums with arbitrary numbers of terms are admitted. In reality one can confine oneself to sums of length $\leq n+1$.

Lemma 1. *Given an arbitrary convex set $M \subset \mathbf{R}^n$ and an arbitrary point $x \in \mathbf{R}^n$, the convex hull of the set $M \cup \{x\}$ is the union of all segments joining x with points of M.*

PROOF. The indicated segments are obviously included in $\operatorname{conv}(M \cup \{x\})$. On the other hand, their union is a convex set. In fact, let y_i be a point on the segment xx_i, where $x_i \in M$, for $i = 1, 2$. Then the segment y_1y_2 is contained in the triangle xx_1x_2, and consequently any of its points belongs to a segment that joins x with some point of the segment x_1x_2, which by the convexity of M belongs to M. □

Lemma 2. *For any set $K \subset \mathbf{R}^n$*

$$\operatorname{conv} K = \Big\{ \sum_{i=1}^{d+1} c_i x_i \mid x_i \in K,\ c_i \geq 0,\ \sum_{i=1}^{d+1} c_i = 1 \Big\},$$

where d is the dimension of the smallest affine subspace containing K.

In other words, $\operatorname{conv} K$ is the union of the d-dimensional simplices with vertices at points of K.

PROOF. It follows from the definition of the convex hull of a set that any of its points belongs to the convex hull of a finite subset of that set. It therefore suffices to prove the assertion of the lemma for finite sets. We proceed by induction on the number of points of the given set. For a one-point set the assertion is obviously true. Suppose that it is true for $L = \{x_1, \ldots, x_m\}$, and let $K = L \cup \{x\}$. Let d (respectively e) be the dimension of the smallest affine subspace that contains K (respectively L). Clearly $d = e$ or $d = e + 1$. We have $\operatorname{conv} L = \bigcup_{i=1}^{s} T_i$, where the T_i are e-dimensional simplices with vertices at the points of L. By Lemma 1,

$$\operatorname{conv} K = \bigcup_{i=1}^{s} \operatorname{conv}(T_i \cup \{x\}).$$

If $d = e + 1$, then $\operatorname{conv}(T_i \cup \{x\})$ is a d-dimensional simplex with vertices at points of K, and the assertion is proved. If $d = e$ and $x \in T_i$, then $\operatorname{conv}(T_i \cup \{x\}) = T_i$. Finally, if $d = e$ and $x \notin T_i$, then every segment xy with $y \in T_i$ intersects a $(d-1)$-dimensional face Γ of T_i, and hence is contained in the union of the d-dimensional simplices T_i and $\operatorname{conv}(\Gamma \cup \{x\})$. □

PROOF of the Theorem. Let $K \subset \mathbf{R}^n$ be a compact set. Let $y_m \in \operatorname{conv} K$ be an arbitrary sequence of points. By Lemma 2, $y_m = \sum_{i=1}^{d+1} c_{mi} x_{mi}$, where $x_{mi} \in K$, $c_{mi} \geq 0$, $\sum_i c_{mi} = 1$, and d denotes the dimension of the smallest affine subspace containing K. Since K is compact and $c_{mi} \in [0,1]$, by passing if necessary to subsequences we may assume that for every i there exist the limits $\lim_{m\to\infty} x_{mi} = x_i \in K$ and $\lim_{m\to\infty} c_{mi} = c_i \in [0,1]$. Passing to the limit we then conclude that $\sum_i c_i = 1$ and

$$\lim_{m\to\infty} y_m = \sum c_i x_i \in \operatorname{conv} K.$$

We have thus proved that from every sequence of points in $\operatorname{conv} K$ one can extract a subsequence that converges to a point of $\operatorname{conv} K$. That is to say, K is compact, as was to be shown. □

A4. Conjugate Elements in Groups

Two elements x, y of the group G are said to be CONJUGATE (written $x \sim y$) if there is a $g \in G$ such that $y = gxg^{-1}$. The set of all elements conjugate to a given $x \in G$ is the orbit of x under the action a of G on itself by inner automorphisms:

$$a(g)x = gxg^{-1}.$$

This shows that *conjugacy is an equivalence relation.*

In the case where G is a group of transformations of some set P, i.e., a subgroup of $S(P)$, the conjugacy of elements of G admits the following intuitive description. Every transformation $x \in G$ can be represented as a diagram consisting of arrows that connect each point $p \in P$ with its image xp. Let $y = gxg^{-1}$. Then

$$ygp = gxp$$

for all $p \in P$. In other words, if x takes p into q, then y takes gp into gq. This means that *the diagram of the tranformation y is obtained from that of x by subjecting all arrows to the transformation g.*

Examples.

1. $G = S_n$. Suppose that the permutation $\sigma \in S_n$ is decomposed into a product of independent cycles:

$$\sigma = (i_1 i_2 \ldots i_s)(j_1 j_2 \ldots j_t) \ldots .$$

Then for each $\gamma \in S_n$ the conjugate permutation $\tau = \gamma\sigma\gamma^{-1}$ decomposes into a product of independent cycles as follows:

$$\tau = (\gamma(i_1)\gamma(i_2)\dots\gamma(i_s))(\gamma(j_1)\gamma(j_2)\dots\gamma(j_t))\dots .$$

In particular, the lengths of the cycles occurring in the decompositions of σ and τ are the same. Conversely, if the permutations σ and τ decompose into products of independent cycles of identical lengths, then they are conjugate. For example, let

$$\sigma = (13)(2456)$$

and

$$\tau = (56)(1432).$$

Then $\tau = \gamma\sigma\gamma^{-1}$, where

$$\gamma = \begin{pmatrix} 1 & 3 & 2 & 4 & 5 & 6 \\ 5 & 6 & 1 & 4 & 3 & 2 \end{pmatrix} = \begin{pmatrix} 1 & 2 & 3 & 4 & 5 & 6 \\ 5 & 1 & 6 & 4 & 3 & 2 \end{pmatrix}.$$

2. $G = \mathrm{O}_3$. Let α denote the rotation around the axis ℓ through an angle ϕ. Then $\beta = \gamma\alpha\gamma^{-1}$, where $\gamma \in \mathrm{O}_3$, is the rotation through the same angle ϕ around the axis $\gamma\ell$. The same holds true for rotary reflections.

Answers and Hints to Exercises

Section 0

1. Use the fact that the map $t \mapsto \det e^{tA}$ is a group homomorphism of $\mathbf{R}$ into $\mathbf{R}^*$.

2. a) $\begin{pmatrix} 0 & 1 \\ 1 & 0 \end{pmatrix}$; b) $\begin{pmatrix} 0 & 1 \\ 0 & 0 \end{pmatrix}$.

5. Infinitely many.

6. Consider the linear operator α given in some basis (e) by the matrix A, and find the matrix of e^α in the basis $(e)C$ in two ways.

7. It is in the cases b), c), and e).

9. a) $n \mapsto \alpha^n$, where α is an arbitrary linear operator;

   b) $n + m\mathbf{Z} \mapsto \alpha^n$, where $\alpha^m = \varepsilon$.

10. a) $x \mapsto x^\alpha$, where α is an arbitrary linear operator;

    b) $z \mapsto z^\alpha$, where α is a diagonalizable linear operator with integer eigenvalues. *Hint.* Consider the composition of the sought-for representation and the homomorphism $\mathbf{R} \to G$ which takes t into e^t in case a) and into e^{it} in case b).

12. Identify the element $x \in \widehat{\mathbf{R}}$ with the line in $\mathbf{R}^2$ passing through the origin of coordinates with slope x^{-1}. Then the transformation $s(A)$ is realized by the linear transformation of the plane $\mathbf{R}^2$ with matrix A.

14. The isomorphism is implemented by the map τ given by $(\tau f)(x) = f(x^{-1})$.

16. Yes.

E.B. Vinberg, *Linear Representations of Groups*, Modern Birkhäuser Classics,
DOI 10.1007/978-3-0348-0063-1, © Birkhäuser Verlag 1989

17. $n + m\mathbf{Z} \mapsto \alpha^n$, where $\alpha^m = \varepsilon$ and the characteristic polynomial of the operator α has real coefficients.

Section 1

2. The space of polynomials of degree $\leq n$ for arbitrary n, the null subspace, and the whole space.

3. Any subspace spanned by some set of eigenvectors of the operator α.

7. a) No; b) Yes.

14. Using first only the invariance under the group of diagonal matrices, show that every invariant subspace that contains nonscalar matrices necessarily contains a "matrix unit" E_{ij} with $i \neq j$.

15. Using first only the invariance under the group of diagonal matrices, show that every nonnull invariant subspace contains a matrix of the form $E_{ij} + E_{ji}$ or $E_{ij} - E_{ji}$, with $i \neq j$.

Section 2

4. $\mathbf{Z}_m$ and $\mathbf{T}$.

Section 3

1. It is the trivial representation in the dual space.

10. $T^2(g)X = T(g)XT(g)'$.

11. $\gamma = \alpha \otimes \varepsilon + \varepsilon \otimes \beta$.

13. Since both sides of equality (9) are linear in ξ, it suffices to verify it for $\xi = v \otimes f$, where $v \in V$, $f \in V'$.

14. $(T \otimes T)(g_1, g_2)X = T(g_1)XT(g_2)'$.

16. All possible sums of the one-dimensional representations I and det.

17. The three representations obtained by lifting the one-dimensional representations of the group $A_4/(A_4, A_4) \simeq \mathbf{Z}_3$.

18. First show that every "elementary" matrix $E + tE_{ij}$ $(i \neq j)$ belongs to the commutator subgroup by verifying that it is the commutator of a diagonal matrix and an elementary matrix (with the same i and j).

Section 4

2. Let α be an endomorphism of the representation $M|_G$. Using the fact that α commutes with the transformations belonging to G which leave the symbol 1 fixed, show first that $\alpha(e_1) = ae_1 + b(e_1 + \ldots + e_n)$ (with $a, b \in \mathbf{C}$).

5. All rotations.

7. Use Theorem 6 of 3.7.

8. Show that a morphism of the representation R into an arbitrary representation T is uniquely determined by the vector into which it takes the function δ_e (see 0.9), and that the latter can be any vector in the representation space of T.

10. For real representations this is false.

11. Use Exercise 10.

12. Show that $R_{\mathrm{M(T)}}$ is a multiple of T for every irreducible representation T.

14. See the hint to Exercise 12.

15. Use the fact that if f_0 and f_1 are two iner products in the real vector space V, then there is a linear operator σ, symmetric with respect to f_0, such that $f(x, y) = f_0(\sigma x, y)$.

Section 5

1.

	ε	(12)	(23)	(13)	(123)	(132)
$T_{1,11}$	1	1	1	1	1	1
$T_{2,11}$	1	-1	-1	-1	1	1
$T_{3,11}$	1	0	0	0	ω	$\bar{\omega}$
$T_{3,12}$	0	1	$\bar{\omega}$	ω	0	0
$T_{3,21}$	0	1	ω	$\bar{\omega}$	0	0
$T_{3,22}$	1	0	0	0	$\bar{\omega}$	ω

$$\left(\omega = -\tfrac{1}{2} + i\tfrac{\sqrt{3}}{2}\right).$$

2. The dimension of the representation.

3. a)

	ε	(12)	(123)
χ_1	1	1	1
χ_2	1	-1	1
χ_3	2	0	-1

b) $\chi_L(g) = \chi_R(g) = \begin{cases} |G| & \text{for } g = e, \\ 0 & \text{for } g \neq e. \end{cases}$

4. Three one-dimensional representations T_1, T_2, T_3 (see Exercise 17, Section 3) and one three-dimensional representation T_4, under which A_4 is mapped onto the group of rotations of a tetrahedron. Their characters are given in the following table:

	ε	(12)(34)	(123)	(132)
χ_1	1	1	1	1
χ_2	1	1	ω	$\bar{\omega}$
χ_3	1	1	$\bar{\omega}$	ω
χ_4	3	-1	0	0

$$\left(\omega = -\tfrac{1}{2} + i\tfrac{\sqrt{3}}{2}\right).$$

7. a) See Example 2 of A1.5;

b) four one-dimensional representations and the $n-1$ two-dimensional representations given by the rule

$$a \mapsto \begin{pmatrix} e^{k\pi i/n} & 0 \\ 0 & e^{-k\pi i/n} \end{pmatrix}, \qquad b \mapsto \begin{pmatrix} 0 & (-1)^k \\ 1 & 0 \end{pmatrix}$$

$(k = 1, \dots, n-1)$.

8. For all $n \neq 3$.

10. a) $I + \Pi + \mathrm{Id} + (\mathrm{Id})\Pi$ (and $\mathrm{Id} \simeq M_0\Pi$);

b) $I + P + \mathrm{Id}$ (and $\mathrm{Id} \simeq M_0$). (Here Π is a nontrivial one-dimensional representation and P an irreducible two-dimensional representation.)

11. See Example 2, Section 6.

Section 6

3.

	ε	(12)	(12)(34)	(123)	(1234)
χ_I	1	1	1	1	1
χ_Π	1	-1	1	1	-1
χ_P	2	0	2	-1	0
χ_{M_0}	3	1	-1	0	-1
$\chi_{M_0\Pi}$	3	-1	-1	0	1
	1	6	3	8	6 ;

$$M_0^2 \simeq I + P + M_0 + M_0\Pi.$$

5. $TT' \simeq T^2 \simeq T_1 + T + T_5$ (see Example 2 in Section 6). The invariant subspaces $U_1, U_2, U_3 \subset \mathrm{L}(V)$ corresponding to the terms of this decomposition can be described respectively as the space of scalar, skew-symmetric, and symmetric traceless operators.

Section 7

1. Show that the group $R(\mathrm{SU}_2 \times \mathrm{SU}_2)$ acts transitively on the unit sphere in the space $\mathbf{H}$ and contains all rotations that leave fixed the matrix E.

2. $\begin{pmatrix} a & b \\ c & -a \end{pmatrix} \mapsto bu_1^2 - 2au_1u_2 - cu_2^2.$

4. $\chi_n(A(z)) = (z^{n+1} - z^{-(n+1)}/(z - z^{-1}).$

Section 8

2. Use the fact that the linear span of the functions $z \mapsto z^n$ is dense in the space of continuous functions on the circle.

3. $(z_1, \dots, z_k) \mapsto z_1^{n_1} \dots z_k^{n_k}$ (with $n_1, \dots, n_k \in \mathbf{Z}$).

6. See 9.1.

8. Establish the corresponding equality for characters.

10. Use the fact that any representation of a compact group is uniquely determined by its character.

Section 9

4. Find first all homogeneous harmonic polynomials of degree one and two.

5. Use the fact that the representation $h(z) \mapsto z^m$ of SO_2 occurs in the representation of SO_2 in the space A_m with multiplicity one (see Lemma 4).

6. Show that $\overline{U}_m$ is a minimal SO_3-invariant subspace.

7. Use integration by parts to show that the polynomial

$$\frac{d^m}{dt^m}[(t^2-1)^m]$$

is orthogonal to all polynomials of degree $< m$.

Section 10

3. $t \mapsto \mathrm{e}^{iat}$ (with $a \in \mathbf{R}$).

4. $(t_1, \dots, t_n) \mapsto \mathrm{e}^{c_1 t_1 + \dots + c_n t_n}$ (with $c_1, \dots, c_n \in \mathbf{C}$).

5. $(d\Phi(\xi)f)(x, y) = -f(\xi x, y) - f(x, \xi y)$ or, in matrix form, $d\Phi(X)Y = -X'Y - YX$.

7. a) $\mathrm{L}_n^0(\mathbf{R})$;

b) the algebra of traceless skew-Hermitian matrices;

c) the algebra of all real triangular matrices.

8. Use the following relation between group commutators: $(h, g) = (g, h)^{-1}$.

11. Write the relation defining an automorphism, and differentiate it with respect to g.

12. $ds_* \left(\begin{pmatrix} \alpha & \beta \\ \gamma & \delta \end{pmatrix} \right) = (\gamma x^2 + (\delta - \alpha)x - \beta)\dfrac{\partial}{\partial x}.$

List of Notations

(As a rule, we give here the notations which have been used elsewhere in the book without explanations.)

$\mathbf{N}$, $\mathbf{Z}$, $\mathbf{Q}$, $\mathbf{R}$, $\mathbf{C}$ the sets of natural, integer, rational, real, and complex numbers (may be regarded as an additive group, ring, or field as appropriate);

$\mathbf{R}^*$, $\mathbf{C}^*$ the multiplicative groups of nonzero real and complex numbers;

$\mathbf{T}$ the multiplicative group of complex numbers of modulus one, or (in Sections 7 and 11) the isomorphic group of matrices $\begin{pmatrix} z & 0 \\ 0 & z^{-1} \end{pmatrix}$ with $z \in \mathbf{C}^*$, $|z| = 1$;

$\mathbf{Z}_m = \mathbf{Z}/m\mathbf{Z}$ the additive group of residues mod m;

C_m the abstract multiplicative cyclic group of order m;

$\mathrm{L}_{m,n}(K)$ the set of all $(m \times n)$-matrices with entries in the ring K (may be regarded as a vector space);

$\mathrm{L}_n(K)$ the set of all square matrices of order n with entries in the ring K (may be regarded as a vector space or as a ring);

$\mathrm{L}_n^0(K)$ the set of all traceless matrices in $\mathrm{L}_n(K)$;

$\mathrm{L}_n^+(K)$, $\mathrm{L}_n^-(K)$ the sets of symmetric and skew-symmetric matrices;

$\mathrm{GL}_n(K)$ the group of invertible matrices;

$\mathrm{SL}_n(K)$ the group of matrices with determinant 1;

O_n, U_n the groups of (real) orthogonal matrices and (complex) unitary matrices;

SO_n, SU_n the groups of orthogonal and unitary matrices with determinant 1;

$\det A$, $\operatorname{tr} A$ the determinant and the trace of the matrix A;

A' the transpose of the matrix A;

E the identity matrix;

E_{ij} the matrix whose (i, j) entry is 1, and whereas the other entries are 0 (a "matrix unit");

V' the dual of the vector space V;

$\mathrm{L}(V) = V \otimes V'$ the space of all linear operators on the vector space V;

$\mathrm{GL}(V)$ the group of invertible linear operators;

$\mathrm{SL}(V)$ the group of linear operators with determinant equal to 1;

$\mathrm{B}^+(V)$, $\mathrm{B}^-(V)$ the spaces of symmetric and skew-symmetric bilinear functions on V;

$\det \alpha$, $\operatorname{tr} \alpha$ the determinant and the trace of the linear operator α;

ε the identity linear operator and, more generally, the identity map of any set;

$\alpha_{(e)}$ the matrix of the linear operator α in a basis $(e) = (e_1, \dots, e_n)$;

$\langle M \rangle$ the linear span of the set M in a vector space, or the subgroup generated by the set M in a group;

$|M|$ the number of elements of the finite set M;

$\simeq$ isomorphism;

δ_{ij} the Kronecker symbol;

S_n the group of all permutations of the set $\{1, 2, \dots, n\}$;

A_n the group of all even permutations;

$\Pi(\sigma)$ the parity of the permutation σ;

I the trivial linear representation (generally speaking, of arbitrary dimension).

References

1. Adams, J. F., *Lectures on Lie Groups*, W. A. Benjamin, New York-Amsterdam, 1969.

2. Curtis, C. W., and I. Reiner, *Representation Theory of Finite Groups and Associative Algebras*, Wiley-Interscience, New York, 1962.

3. Kirillov, A. A., *Elements of the Theory of Representations* (Russian), Nauka, Moscow, 1972; English transl.: Springer, Heidelberg-Berlin-New York, 1976.

4. Kostrikin, A. I., *Introduction to Algebra* (Russian), Nauka, Moscow, 1977; English transl.: Springer, New York-Berlin, 1982.

5. Lang, S., *Algebra*, Addison-Wesley, Reading, MA, 1965.

6. Naimark, M. A., *Theory of Representations* (Russian), Nauka, Moscow, 1976; English transl. (with A. I. Shtern, coauthor): Springer, New York-Heidelberg-Berlin, 1982.

7. Pontryagin, L. S., *Continuous Groups* (Russian), 4th ed., Nauka, Moscow, 1984; English transl. of 2nd ed.: *Topological Groups*, Gordon & Breach, New York, 1966.

8. Proskuryakov, I. V., *Collection of Problems in Linear Algebra* (Russian), Nauka, Moscow, 1984; English transl.: Mir, Moscow, 1985.

9. Vilenkin, N. Ya., *Special Functions and the Theory of Group Representations* (Russian), Nauka, Moscow, 1965; English transl.: Translations of Math. Monographs **22**, Amer. Math. Soc., Providence, RI, 1968.

10. Weil, A., *L'Intégration dans les Groupes Topologiques et ses Applications*, Hermann, Paris, 1951.

11. Zhelobenko, D. P., *Compact Lie Groups and Their Representations* (Russian), Nauka, Moscow, 1970; English transl.: Translations of Math. Monographs **40**, Amer. Math. Soc., Providence, RI, 1973.

12. Zhelobenko, D. P., and A. I. Shtern, *Representations of Lie Groups* (Russian), Nauka, Moscow, 1983.

Index

Made in the USA
Monee, IL
13 September 2026

62975225R00090